Women in the Housing Service

This book traces the involvement of women in housing management from 1912 to the present day. Marion Brion shows how the levels of participation fluctuated during this period and looks at factors which have helped or hindered women's involvement in housing. These include changes in policy and practice, strategies of organisation and patterns of employment. A range of techniques, including unique oral sources as well as unpublished documentary material, are used to unveil a previously little researched area.

Throughout, the issues which emerge are related back to the modern context. The separatism which developed in the shape of the Society of Women Housing Managers early this century is compared with the more diverse organisational forms used to support women during the 1980s. Equal opportunities initiatives in housing during the last two decades are compared with those in accountancy, planning, surveying and architecture in what is the first comparative study of this kind.

Breaking new ground with the research presented, *Women in the Housing Service* is an important addition to the literature on equal opportunities and housing. It will be of great interest to students and researchers in history, women's studies and housing.

Marion Brion has been a Housing Manager, Development Officer, Senior Research Fellow and Senior Lecturer, and has been a major contributor to research and practice on education and training for housing work. She currently trains housing staff in interpersonal skills as well as doing research.

Women in the Housing Service

Marion Brion

London and New York

First published 1995
by Routledge
11 New Fetter Lane, London EC4P 4EE

Simultaneously published in the USA and Canada
by Routledge
29 West 35th Street, New York, NY 10001

Typeset in Times by
J&L Composition Ltd, Filey, North Yorkshire
Printed and bound in Great Britain by
Biddles Ltd, Guildford and King's Lynn

British Library Cataloguing in Publication Data
A catalogue record for this book is available from the British Library

ISBN 0–415–08094–0

Library of Congress Cataloging in Publication Data
Brion, Marion.
    Women in the housing service / Marion Brion.
        p.   cm.
    Includes bibliographical references and index.
    ISBN 0–415–08094–0
    1. Housing authorities — Great Britain — History.   2. Housing
management — Great Britain — History.   3. Women in housing management — Great Britain — History.   I. Title.
HD7333.A3B759   1995
363.5′85′082 – dc20   94-7747
                                                                      CIP

Contents

Tables and figures

TABLES

FIGURES

Acknowledgements

I would like to thank the many people who helped with the writing of this book. I am particularly grateful to Rosalie Silverstone who supervised my PhD thesis, providing not only a high level of professional advice but also the encouragement which is so vital. Invaluable practical help in word processing, layout and administration was given by Kelsey Blundell, despite her own increasing disability. Anthea Tinker, my co-author for the earlier work on women and housing, provided a key impetus for the study and a source of help and inspiration along the way.

The study would not have been possible without the initial help of Marjorie Cleaver, then at the Housing Centre Trust, who generously provided unique information on the location of the Society of Housing Managers' members and records. Institute of Housing staff and staff of the RTPI, RICS, RIBA, ICA, Housing Corporation and Housing Centre Trust provided vital information and statistics. I would particularly like to thank the ex-members of the Society of Housing Managers and the Institute of Housing with whom lengthy interviews were carried out. For the later chapters I have benefited from stimulating and varied discussions with women's groups, especially at the National Federation of Housing Associations Women's Conference and Women's Standing Group and the Older Feminists' Network. I hope that, whether they agree with my conclusions or not, they will find something of benefit to them in this book.

Thanks are due to the City University for providing me with the opportunity to begin this study and to librarians there and at the Fawcett Library, London University Institute of Education, Tottenham College, British Library and the Public Record Office for their helpful advice. Finally I am very grateful for the much needed support of friends and members of my family over the years in which I completed this work.

Abbreviations

AGM	annual general meeting
AMA	Association of Municipal Authorities
AMC	Association of Municipal Corporations
AWHPM	Association of Women House Property Managers
AWHW	Association of Women Housing Workers
BTEC	Business and Technology Education Council
CHAC	Central Housing Advisory Committee
CHAR	Campaign for the Homeless and Rootless
CIS	Counter Information Services
Copec	Conference on Christian Politics, Economics and Citizenship
CRE	Commission for Racial Equality
CSI	Chartered Surveyors' Institution
DLO	Direct Labour Organisation
EOC	Equal Opportunities Commission
FBHO	Federation of Black Housing Organisations
GLC	Greater London Council
HERA	Housing Employment Register and Advice
ICA	Institute of Chartered Accountants
IHA	Institute of Housing Administration
IOH	Institute of Housing
JCT	Joint Contracts Tribunal
LCC	London County Council
LGTB	Local Government Training Board

LSE	London School of Economics
MAP	Ministry of Aircraft Production
MHLG	Ministry of Housing and Local Government
MOS	Ministry of Supply
MSC	Manpower Services Commission
NALGO	National Association of Local Government Offices
NCC	National Consumer Council
NCCL	National Council for Civil Liberties
NCVQ	National Council for Vocational Qualifications
NEDO	National Economic Development Office
NFHA	National Federation of Housing Associations
NHTPC	National Housing and Town Planning Council
OPM	Office for Public Management
PEP	Political and Economic Planning
PO	Principal Officer
RIBA	Royal Institute of British Architects
RICS	Royal Institute of Chartered Surveyors
RIPA	Royal Institute of Public Administration
RTPI	Royal Town Planning Institute
SHM	Society of Housing Managers
SWHEM	Society of Women Housing Estate Managers
SWHPM	Society of Women House Property Managers
SWHM	Society of Women Housing Managers
WAHC	Women's Advisory Housing Council
WEB	Women and Built Environment
WHC	Women into Construction Action Group

1 Introduction

IDEALLY WOMEN'S WORK?

It was a woman, Octavia Hill, who first raised awareness of housing management at the end of the nineteenth century. From 1912 women who had worked with her were beginning to form their own organisations. A separate women housing managers' society existed from 1932 to 1948, doing pioneering work and exerting a considerable influence. From 1948 men were able to become members, but women remained in the majority.

This Society of Housing Managers amalgamated with the Institute of Housing (mostly men) in 1965. Women's participation in the Council of the new Institute reduced rapidly until by the early 1970s there was only one woman on the Council. The National Federation of Housing Associations' Council followed a similar pattern. It seemed that there were few women prominent in housing employment.

Was the disappearance of women's influence from housing work in the early 1970s solely due to the loss of their own professional society or were there other factors at work? This was the impetus for an initial study, published in 1980 (Brion and Tinker) which began to identify a little of the history of the Society and the complexity of the factors at work. But this raised as many questions as it answered. Housing work had never fitted neatly into the stereotype of 'fit work for women'. How had women managed not only to establish themselves but also to maintain an influential presence in the period 1930 to 1960? And might they be able to fight back and improve their position again at the end of the century?

The incentive for the research on which this book is based was therefore to identify the factors which had influenced women's employment in housing at the various stages, to record the contribution of the Society of Housing Managers to this and to examine the implications for women in housing work today and in the future. The urgency of present action to secure some of the historical records was also an incentive. In 1970 very little was known about the location of any records of the Society of Housing Managers. Fortunately the ex-deputy secretary of the Institute of Housing (Marjorie Cleaver) mentioned that the bound volumes of the Society had been stored in the basement of the Institute's offices and from her description it was possible to locate these. Some interviews with senior Society members had been carried out in 1979 and had proved very fruitful. By 1980 this generation were in their eighties and nineties and they and any remaining documents they held might be lost, as has so often happened with women's history (Bridenthal, Koonz and Stuard, 1987).

METHODS AND PERSPECTIVES

This book examines different theoretical approaches rather than offering one particular interpretation throughout. It draws methods and theories from three different disciplines: history, psychology and sociology.

Traditionally, disciplines are defined as differing in the concepts that organise experience, the way in which a statement is thought to be true or false, techniques and methods of setting about an enquiry, and range of problems tackled (Pring, 1978). While much lip service is paid to the need for interdisciplinary study, its practice is dangerously vulnerable to charges of eclecticism or superficiality (Riggs, 1990; McCall, 1990). Sociologists have been known to react adversely to the introduction of psychological theory and the relationship between sociology and history has been subject to much debate, sometimes acrimonious. However, the reality of much applied work is that the world is not divided neatly into compartments. Previous experience of housing and educational research had confirmed the author's views that the benefits of a interdisciplinary approach were often worth the risks. To research the early stages of women's involvement in housing required historical method. Comparing and interpreting the findings of modern surveys on women's employment required a firm grasp of survey methodology and broad social trends. Illuminating explanations of the problems experienced by women in the Institute in the late 1960s came from social psychology and psychology. It seemed that different theories were not necessarily in conflict but dealing with different levels. However, the interdisciplinary approach sometimes produced hazards.

For example, a key issue which runs throughout the book is the way in which different occupations are allocated to men or women. Sociologists use the word 'gender' to describe the social construction of men's and women's roles and there are good arguments that this is the most accurate word (see, for example, Garrett, 1987). But the majority of psychologists in the period under study used the phrase 'sex role stereotyping'. This phrase has also been used when discussing this type of research.

Another issue is that oral history techniques (which contributed important data to this study) have most frequently been used in the history of groups excluded from the power elite, such as women, because the record of women's achievement is often submerged (Beddoe, 1983). Dubois (1980) regards the study of elite women's groups as a rather rudimentary stage in feminist history and she is not alone in this. But the Society of Housing Managers was an elite group mainly in relation to other women; the book records their struggle in relation to the male dominated world. Included in Chapters 8 to 11 is a description of the way in which the record of their achievement was being submerged. It is to be hoped that the reappraisal of feminism taking place in the 1990s will enable us to appreciate that it is important to continue to set the (mainly male generated) record straight about the achievements of women in the past as well as about the lives of others who are too often unrecorded. For example, Gianna Pomata (1993) has recently argued the case for 'particular' history which recognises its limitations, rather than falsely claiming universality. In the present study professional records were available for the earliest period but it is only the more modern surveys which cast some light on the many women working at non-professional levels in housing. Where such evidence is available it is used.

Defining the scope of the enquiry

This study covers 80 years of women's employment in housing. As Cohen and Manion (1985) state, historical subjects inevitably produce acute problems in defining the area of enquiry, since each topic has many related or contingent areas which seem worth exploring. In order to place the developments in women's employment in context it was necessary, for each historical period, to identify what changes were taking place in housing policy and administration. While policy was, for most periods, covered in standard texts, this was not true of administration, which had been neglected. Coverage of the Second

World War period was extremely sparse. For the thesis on which this work is based (Brion, 1989) original research was undertaken to piece together sufficient background material. This has been summarised very briefly in this book. Readers interested in the detail of housing administration, especially of the war period, are advised to consult the original thesis or related articles.

The other major contingent area was feminism, which was important both for the historical period and the present. There is now extensive work on women pre-1939 which could only be used very selectively. Basic secondary sources dealing with women's employment have been used at appropriate stages. It was clear that in the 1920s and 1930s the Society had extensive links with other campaigning organisations but these have not been pursued as they would have unduly extended the area of the study. Feminist theories are drawn on more extensively in the latter part of the book but it was still necessary to be highly selective.

The Society of Housing Managers' minutes, annual reports and journal provided a major primary source. These were not all together in one place but the Fawcett Library had a reasonable collection of Society publications, though not complete. The British Library Index contained few references to the Society of Housing Managers, though it was a good source of data for those periods when housing management had been the subject of official committees and for some aspects of the war period. Given the limitations of the published sources it was hoped that private ones would be more fruitful. However, many of the older women housing managers had moved a number of times and unsentimentally thrown out earlier documents. Fortunately Miss Janet Upcott, who had worked with Octavia Hill at the end of her life and had continued in housing management for many years, had kept material from the women's societies of the 1920s and 1930s. She donated a unique collection of Society of Housing Managers' reports and journals and associated material to the author. Some other interviewees also passed on relevant documents.

The more modern period presented a different set of issues. Statistical material was available in the Institute of Housing records. These had not been analysed by gender so, for 1965, 1977 and 1983, this had to be done manually, which was very time consuming. Surveys which produced data about men's and women's employment in housing were carried out by the City University, the Institute of Housing and the NFHA but produced problems about comparisons. Special studies were carried out by the author in 1984 and 1992 to gather the data about women's employment in other professions.

Official reports and committee minutes were also used for this period.

The interviews – choice of interviewees and achieved sample

The initial group chosen for interview were the members of the Standing Joint Committee which dealt with the unification of the Society and the Institute. In tracing ex-Society members, many of whom had left housing before the 1970s, the social contacts formed by the Society were very helpful. The Institute members had also been people of some standing in housing and it was possible to trace them. Where interviews were not achieved it was because of the death or serious illness of the member, except in one case where the person concerned felt that his memory was too poor for an interview but did give some comments over the telephone. Nevertheless, the impact of male mortality meant that fewer of the Institute members survived to be interviewed, so the achieved interviews are weighted towards the Society. From this initial round of interviews a list of key figures was made and this was added to as the interviews progressed. Priority was given to interviewing those mentioned by more than one person. Mainly this second group were the older generation who had not been actively involved in unification. In all, 29 interviews were achieved.

The interview method

The interviews were usually carried out in subjects' homes and required very careful preparation. Many elderly people are wary of being approached by strangers, and people who have been in senior positions in housing organisations are often cautious about making statements to the press or writers. The social network and the author's past membership of the Society provided a source of reference which helped to open doors. The interviews were carried out in a semi-structured format using a prompt sheet which, if interviewees wished, was sent to them in advance (see Appendix). Care was taken to phrase questions as neutrally as possible and to keep interviewer statements to a minimum, with a concentration on active listening (see, for example, Banaka, 1971, for a discussion of the skills required for in-depth interviewing). Because of the age of the interviewees a very flexible attitude was adopted to the schedule and breaks were taken in the interviews which were often quite lengthy. All interviews were tape-recorded where possible and only one respondent was not happy with this. It has sometimes been argued that respondents may find tape recording or even note taking off-putting. These interviewees, with their businesslike backgrounds, usually seemed to appreciate the need to take notes as it confirmed that their views were being taken seriously. They sometimes did make interesting comments later, 'off the record', but this would be when both tape recorder and notebook were put away. Such comments were taken on board in a general sense to respect the confidentiality.

Most interviewees were concerned about confidentiality; this was assured in writing up and making comments anonymous, and where absolutely necessary omitting details which might have identified the respondent. This was less than ideal with regard to the archival nature of the material but it was the only viable way to carry out research with this particular group. The interviews were transcribed and analysed by theme using a card index and colour coding – a very lengthy process but essential in this kind of qualitative research (Lummis, 1987; Burgess, 1985).

Uses and limitations of the interviews as a data source

The interviews asked respondents to look back to events that happened twenty, thirty or more years ago and human memory is known to be fallible. However, it is known that events with emotional significance are more vividly recalled (Rose, 1985: 63) so the use of interviews as a way of illustrating what people felt at the time is valid. In some instances interviews were used as sources of factual material which is not well covered by the literature. This was particularly so in the interviews with Miss Upcott, for example. Then over 80, she was one surviving link with the managers who had worked with Octavia Hill. She was able to cast light on the events between Octavia Hill's death and 1932 which were only covered briefly and in a fragmentary way in the literature. Where Miss Upcott's account was checked against written records its accuracy was borne out – when she was uncertain about dates she said so.

Balance and bias in the sources and methods used

Samuel (1980: 165) traced the desire to use oral history methods to roots as diverse as a reaction against some of the conventions of academic history, the rise of the women's movement and Marxist theory. He argues that the problems of reconstructing and interpreting the past are common to the use of both oral and written sources.

We are continually aware, in our work, of the silences of the record; while in writing we know how the addition of a single word can transform the thrust of a sentence. No historian can doubt the fragility of the construction of historical knowledge, or be unaware of the filters and selectivities which intervene between the original document and the reproduction of some fragment of it in the pages of a finished work.

(Samuel, 1980: 175)

The silences of the record are evident in various ways. For example, weeding of the records at the Public Record Office has an effect in the restricted number of documents relating to the war period, which seem largely to be those used by Titmuss (1950). Some of the committees on housing management have very full records; some have very little. However, there is supplementary material available in the professional and local authority journals, though it is often fragmented. In the interviews the majority of informants were members of the Society. For matters internal to the Society this is often helpful but it is clearly of more importance when matters like the influence of the Society are being discussed and here efforts have been made to seek external sources of information and statistical data.

This study uses interviews, written records and survey material as well as secondary data, sometimes to illuminate different issues but sometimes in relation to the same one. The use of two or more approaches may be of substantial benefit. Cohen and Manion use the word triangulation to describe this (Cohen and Manion, 1985: 208): 'The fact that different sets of data point to similar conclusions gives one more confidence in the interpretation though one must be aware that, particularly when causation is being discussed, what is being produced is the ''best fit'' between the theories so far available and the data so far available.'

Portelli's view that 'The first thing that makes oral history different, therefore, is that it tells us less about their events than about their meaning' (Portelli, 1981: 99) is also relevant. The recognition of personal and subjective elements is an important theme in this book. For example, in explaining the reasons why women's participation in the Institute dropped after 1965 the subjective experience of the interviewees was particularly useful; it could then be related to social and psychological theories.

Awareness of the importance of the meaning which both subjects and researchers give to events has fortunately been reinstated in historical research (for example Samuel, 1980; Smith, 1989) and in at least some psychological and sociological approaches (Hartnett, Boden and Fuller, 1979; Erikson, 1973). All existing accounts contain bias. Feminists and minority groups have often pointed out that where bias is that of the dominant group it is often not made explicit or noticed (Waring, 1989; Coote and Campbell, 1982). In this context the fact that the author was herself at one time a member of the Society and is an older, educated white woman, a long-term resident in a multiracial area, adds a level of understanding but also sources of bias. However, this is a spur to following good practice in identifying sources of data, making the arguments explicit and leaving readers to make up their own minds.

While awareness of bias and of the influence of different frameworks colours the whole study, the 1980s posed particular problems because of the author's own involvement in many of the events described. Since it was more difficult to achieve a balanced appraisal of events so recent, copies of the drafts were sent to a range of individuals and organisations involved for consultation. Some very helpful comments were received and these are reflected in the text.

OUTLINE OF THE STUDY

Women's work in housing began with Octavia Hill, who is relatively well known. Critiques of her work are examined in Chapter 2. Has she been treated less fairly than female pioneers in other occupations and if so, why?

From 1912 onwards women who had been trained by Octavia Hill began to form organised groups. By 1932 these groups had joined together in one Society for trained women working in housing management. Why were they attracted to work in housing and why did they face opposition? In Chapter 3 the opposition encountered by these women is examined in the light of theories of gender and stereotyping. The efforts made by the women and their supporters to counteract the opposition are described.

Chapters 4 and 5 describe the growth of social housing in the inter-war period and the employment opportunities this offered to women. The important role which the women's Society played in negotiations with employers and in promoting training and social support for women managers is identified and analysed. The growth of the Institute of Housing and its relationship with the Society is described. Was there any substantial difference in the approach of the two bodies to professionalism and to the practice of housing management? This question is discussed with particular reference to the functioning of the Balfour Committee and to primary and secondary data.

The Second World War period is almost ignored by all the existing histories of housing. Was this because the focus was on management rather than on new building? Or is it because this is a period when women played a most crucial role? Chapter 6 discusses the work of women housing managers during the war with particular reference to the 'reserve army of labour' theory. Some interesting developments in the role of the state in housing and in the nature of housing management are identified.

In 1948 the Society admitted men to membership and in 1965 it amalgamated with the Institute of Housing. Chapter 7 identifies the factors in post-war expansion which disadvantaged women and describes the events which led up to unification between the Society and the Institute. After this the book begins to depart from the narrative structure.

Chapter 8 looks at women's participation in the Institute in the years 1965 to 1984 and draws on the experience of interviewees to identify why women were adversely affected by the amalgamation. Explanations relating to group dynamics, gender, stereotyping and power are examined. Chapter 9 takes a much broader view and looks at the fortunes of women in housing employment against a background of social change and change in housing organisations. At this point it becomes possible to extend the discussion beyond the professional level of employment. Chapter 10 extends the scope to women in four related professions: planning, architecture, surveying and accountancy. What are the reasons why women have been able to make more gains in accountancy than in the occupations closest to the building industry? Is this situation likely to change and what significance does it have for women in housing employment?

Chapter 11 ends the narrative by considering the 'fight back' by women in housing employment in the 1980s. It describes how the discussion was extended to the concerns of women in non-professional jobs and to women in their role as users of housing. A brief comparison is made with the housing profession's response to race equality issues.

Chapter 12 reviews the evidence on key issues. How far is separate organisation essential to maintain women's participation in any occupation attractive to men? What are the specific factors which help or hinder women's participation in housing? What seem to be useful tactics and strategies for campaigning? The significance of these findings for wider debates about women's place in society is considered.

2 Octavia Hill

It is impossible to write about the Society of Housing Managers without some consideration of the influence of Octavia Hill. This chapter begins with a brief discussion of her housing work and the 'Octavia Hill system' but the main focus is on the different ways in which Octavia Hill's work has been interpreted. Three main questions are raised. How extensive and substantial was Octavia Hill's influence on housing management? What exactly was the nature of that influence? Why does Octavia Hill seem to arouse so much controversy and strong feeling?

OCTAVIA HILL'S HOUSING WORK

Octavia Hill's grandfather, Southwood Smith, was prominent in sanitary reform in the early years of the nineteenth century (Moberly Bell, 1942: 5). At the age of 14 Octavia Hill started work at a ragged school and became aware of the poor housing conditions of her pupils. The movement for sanitary reform of housing was gaining strength in the early 1840s and 1850s and a number of reform bodies associated with housing were founded, including Model Dwellings companies. An account of the first meeting of the Association for Sanitary Reform in 1859 gives one of the first formal records of women's role being stressed in this context.

> Lord Shaftesbury, the Chairman, explained how much of the work in its practical detail was specially suited to women, while the legislative must be done by men. Kingsley . . . dwelt on the high infantile mortality and inexorable fate that hung over so many babies. He urged the necessity for women to take up the work because on it the saving of infant life so much depended.
>
> (Tabor, 1927:12)

We can see that these early calls for the involvement of women in housing were based very much on the stereotyped view of woman as the 'ministering angel' common in Victorian charitable and evangelical literature (see for example Wohl, 1977: 184). A few years later, in 1865, Octavia Hill, with Ruskin's help, converted this idea into reality. Octavia was looking for housing for some of the pupils she was concerned with. Ruskin provided the capital to buy three houses but, consistent with the ideas of 'five per cent philanthropy', Octavia was to manage the houses to produce a 'fair percentage' on the capital invested (Hill, 1875: 16).

From the beginning, in managing these properties Octavia Hill put into practice a careful attention to the landlord's duties, such as repair, together with a personal relationship with the tenants and an expectation that in the long run the tenants would behave in a way which was responsible both to their neighbours and to the landlord (for example Hill, 1875: 16–19). From this small start her reputation for managing property began to spread and other people began to hand over properties to manage or money to invest in such property.

One reason for this reputation was that from an early stage Octavia Hill began to write about her work. The writing was vivid and descriptive with plenty of anecdotes and was very positive about what could be achieved. Thus her views became widely known.

As the work grew Octavia Hill drew in other workers. First of all these were lady volunteers, though Octavia Hill always chose and trained them carefully, but later she began to see the need for more systematic recruitment of workers.

Initially, most of the properties concerned were in the West End of London. An attempt to expand work in Deptford was not successful. In Southwark the work started with properties belonging to private owners. The Ecclesiastical Commissioners as ground landlords helped private individuals to purchase building leases and in 1884 handed over to Octavia a group of old courts to manage. On being satisfied with this experiment, they began to hand over to her the management of larger blocks of working-class property, though the Commissioners retained control of capital expenditure (Upcott, 1923: 18–22). The involvement with an 'Institutional Landlord' was an important expansion of the work and itself doubtless helped to pave the way for the employment, in 1916 after Octavia Hill's death, of Octavia Hill trained managers by the Commissioners of Woods and Forests (later the Crown Estate Commissioners).

The work continued to expand slowly, yet the numbers of houses managed was limited. Miss Jeffery estimated that by 1912 (the year of her death) Octavia Hill directly controlled between 1,800 and 1,900 houses and flats exclusive of rooms in tenement houses, although managers trained by her controlled other properties (Jeffery, 1929: 1). Power (1987: 14) states that 'she must have controlled or influenced the management of about 15,000 properties' with about 50 trained women managers working with her. The basis for the latter estimate is not, however, stated.

What was the Octavia Hill system?

One of the central tasks of this study is to trace the way in which Octavia Hill's work led to the formation of the Society of Housing Managers and how the Society of Housing Managers then supported women's employment in housing. Women from the Society tended to become identified with a particular type of management known as the Octavia Hill system and this influenced attitudes towards them. At this stage it is therefore important to look more closely at what Octavia Hill's own ideas about housing management were.

First of all it is important to note that Octavia Hill was a doer and not a theorist. 'I have been asked to add a few words about the houses of the people, but what can I say? There has been so much said. Is it not better now just silently to do?' (Hill, 1884: 11). She did not want to lay down any rigid system or even found an association.

> My friends know that it has never seemed to me well to form any association, whether of the owners of the various groups of dwellings we manage or of ourselves, the workers, who manage . . . Societies, cannot create a spirit . . . All we can do is, where the spirit exists, to try to qualify workers by giving them training, and then link them with owner and group of tenants.
>
> (Hill, 1889)

Octavia Hill stressed that housing management should be an integrated function and this is what became known as the Octavia Hill system. Octavia Hill housing managers collected the rent, dealt directly with problems concerning the tenants and with repairs (at least with small routine repairs). Much later criticism centred around the issue of whether this system was suitable for the changed circumstances of modern housing. As we have seen, Octavia Hill was emphatic about the need to avoid a rigid way of working. Later chapters of this study will discuss the extent to which the Octavia Hill system was advocated by the Society of Housing Managers and the effect which this may have had on women's work in housing.

The emphasis on the value of human relationships is consistent in Octavia Hill's work but, unfortunately, from the point of view of modern commentators is linked with an opposition to state intervention and a tendency towards authoritarianism. Miss Upcott, who had worked with Octavia Hill, commented that of course the local authorities which Octavia Hill knew were not like the present-day ones and the previous corruption of such bodies coloured her views of them. But it is incontrovertible that the substantial gains in housing standards for working-class people eventually came from state financial support for local authority building and that Octavia Hill was opposed to this. She also did have firm ideas about the poor as people who needed to be 'trained and guided' and, in her relationship with the work and her own staff, could be autocratic, though she also made considerable efforts to delegate and to develop and support the staff she delegated to.

The training of workers for housing

Octavia Hill had started off by using lady volunteers but, in her view, such volunteers needed to be trained. However, by the time her work had expanded, she realised that paid workers would be needed and would have to be trained.

In practice the Octavia Hill training required both dedication and hard work. Miss Upcott, who was taken on by Octavia Hill around 1912 after achieving a degree at Somerville and social work training at LSE (Dennison House), recalled the offer of her first post. She was told

> ' "Because we couldn't trust you with any decisions at the moment we offer you £70 per year. But you will have to live over the office and pay £30 a year in rent." So I netted £40; of course I was a trainee . . . I didn't know anything about housing . . . that was the last year of her life; very good for me because she really did train one, she was very fine. Very alarming, you know; you knew you were in the presence of someone rather great . . . The reason I had to live over the office was that Miss Hill managed properties all over London and the managers used to bring their books up to her on a Thursday.'

(Upcott, 1979)

The contribution of Octavia Hill to training is acknowledged by many modern writers, but they are often considerably more critical about other aspects of her work. The next section contrasts the views of the earlier writers about Octavia Hill with those of the critics of the 1970s and 1980s.

ASSESSMENT OF OCTAVIA HILL'S WORK

Early biographers and commentators regarded Octavia Hill with great admiration. 'Octavia Hill and Florence Nightingale were the two greatest women of the nineteenth century.' Thus W. T. Hill, whose biography was published in 1956, quotes Lionel Curtis, first honorary secretary of the National Trust (Hill, 1956: 193). He also quotes the phrase 'a sainted name', attributed to Arthur Greenwood when Minister of Health (Hill, 1956: 191). Moberley Bell's earlier and probably better known biography also pays glowing tributes. 'I am fully convinced that the best of all that is now being done for the better housing of the poor has had for its origin and inspiration the life-work of this remarkable Englishwoman.' (Moberley Bell, 1942)

In contrast to Moberley Bell's admiration, Beatrice Webb is often quoted as an early critic. She felt that Octavia Hill's approach was undermined by her lack of understanding of the true nature of poverty:

> 'The lady collectors are an altogether superficial thing. Undoubtedly their gentleness and kindness brings light into many homes; but what are they in the face of this collective brutality, heaped up together in infectious contact; adding to each other's dirt, physical and moral'.
>
> (Webb, quoted by Wohl, 1977: 189)

But Beatrice Webb shared some of Octavia's views.

> Beatrice Webb, with the typical self-confidence of her class, bordering on arrogance, wrote that she had few misgivings about intruding upon the privacy of the poorer classes: 'rents had to be collected, and it seemed to me, on balance, advantageous to the tenants of low-class property to have to pay their money to persons of intelligence and goodwill'.
>
> (Wohl, 1977: 187)

Tarn and Wohl

Tarn (1973) also views Octavia Hill's approach as superficial: 'Whatever her initial success, the solution was never more than a palliative'. Tarn also considers Octavia 'reactionary as far as physical realities were concerned . . . she did not take a long-term view of the problems and gave her support to immediate action'.

Wohl, on the other hand, balances his criticism by demonstrating how much Octavia Hill cared about the physical surroundings in which working-class people lived. 'Her whole being revolted against the stark and barren working-class blocks which model dwelling companies were putting up throughout London' (Wohl, 1977: 192).

In fact, Wohl gives one of the most extensive and well thought out appraisals of Octavia Hill's work. He starts with a fair summary of the existing situation:

> Of the mid-Victorian housing reformers, Octavia Hill was by far the most widely known and respected. Yet it has been her fate to survive, rather like some great 'classic', well-known by name, but neglected and unread, and she remains the most misunderstood and inadequately handled of the major Victorian reformers. She has, unfortunately, been the victim of partisan history.
>
> (Wohl, 1977: 179)

Wohl reviews her work, its limited scope, and admits 'one must stress at the outset that, unlike nearly all the others in the field, Octavia Hill managed to reach the less prosperous and irregularly employed labouring classes' (Wohl, 1977: 184). He criticises Octavia Hill's 'despotism' though he acknowledges the beneficial effects her influence could sometimes have. Wohl's strongest criticism was of her underlying philosophy of housing:

> Her outstanding failure was that, in the decades when her contemporaries grasped the essence of the housing question to be one of supply and demand, Miss Hill plodded patiently forward, blithely patching up the few houses under her control, almost glorying in petty detail.
>
> (Wohl, 1977: 195)

Wohl's lengthy assessment ends with some fairly stringent criticism. Two more commentators of the 1980s were considerably more acerbic.

Malpass, for example, does not mince words. He starts off by saying 'Most writers on housing management still refer to her pioneering work, often implying that she established modern housing management principles almost single-handed' (Malpass, 1982: 206).

He summarises her approach:

To Hill, housing management meant patient and firm education of the poor in how to lead better lives, as defined by the values of the middle and upper classes. This distinguished her approach from both the commercial management style of the up-market capitalist landlord, and the chaotic non-management style of the down-market slum owners.

(Malpass, 1982: 207)

Like Wohl, Malpass stresses both Octavia's despotism and her opposition to state intervention. He acknowledges her contribution but emphasises limitations:

She had a genuine claim to be one of the pioneers of both housing management and social work. In particular, her emphasis on establishing close relationships with individuals and families in need has formed the basis of a continuing tradition in social work and retains its relevance in housing.

(Malpass, 1982: 229)

However, the form of housing management which she devised has played only a minor part in the development of modern practice. Whereas she opposed state intervention and relied on women volunteers to work closely with tenants, it is council housing, run in bureaucratic fashion by a salaried professional group dominated by men, which has become the main setting for the management of rented housing . . .

Despite Octavia Hill's valuable pioneering work in this field, housing management was in effect reinvented in the 1920s as a wholly administrative activity centred on local government and lacking the moralistic overtones of her method.

(Malpass, 1982: 208)

Spicker was even more stringent in his criticisms of Octavia Hill and those who support her views.

Octavia Hill is clearly held in great regard. An article in *Voluntary Housing* outlines her approach and exhorts us to 'Go back to the origins of management.' . . . A fancied slight to her memory attracts two pages of letters in *Housing* . . . But the legacy she left to housing managers has been baneful. She founded a tradition which is inconsistent with the rights of tenants and destructive of their welfare. Octavia Hill's practice was based in the Christian Socialism of the 19th century; her writings drip with Victorian piety. The positive side of this doctrine was an idealistic belief in the universal rights of all people, including the lower-classes. The negative side was a highly moralistic and judgmental view of her tenants.

(Spicker, 1985: 39)

The emotive language in which this article is written is itself of some interest in the current context. Spicker acknowledged Octavia Hill's positive aims but has a poor view of the effects of her work. He argues that certain established practices in housing management which, 'although they are dying out' are still practised in some places, result from the influence of Octavia Hill and her disciples. These practices are: the exaggerated emphasis on rent payment; the use of notice and eviction as a primary sanction against tenants; the emphasis on cleanliness; and the treatment of 'unsatisfactory' tenants' and 'problem families' – treatment firmly rooted in a pathological view of poverty. He acknowledges Octavia Hill's contribution to ideas of generic housing management and decentralisation but concludes,

> It would be unfair to attack Octavia Hill too harshly for her principles; she was, after all, a woman of her time. But this is not to say that the same principles can be accepted in the present day.

(Spicker, 1985: 40)

In contrast to Malpass and Spicker, Boyd, also writing in the 1980s, gives a more sympathetic view, but from a very different standpoint. Boyd discusses Octavia Hill and two other prominent Victorian women, Josephine Butler and Florence Nightingale: 'England's three great nineteenth century pioneers of social reform' (Boyd, 1982: xi). Because Boyd's major interest is in the theological views of these women, she brings a different perspective. Her discussion focuses particularly on Florence Nightingale, considering why she is better known than the other two.

> The legend of Florence Nightingale contained much that people wanted to hear over and over again. It centred on two folk heroes – the British soldier and the woman who serves him. It shows each in a noble light. Furthermore, it epitomised what the Victorians believed to be the ideal relationship between man and woman . . .
> If the legend brought reassurance to the Victorian male – and encouragement to the common soldier whose qualities had long been unappreciated – it also brought hope to the Victorian middle-class woman. Without disturbing the underlying assumptions in

the male–female relationship, it showed a woman, living in a setting of danger and excitement, making important decisions, taking on important responsibilities.

(Boyd, 1982: 186, 187)

Boyd comments on the way that each of these leaders remained ambivalent in her attitude to feminist issues. Feminists may find Octavia Hill's opposition to the extension of the suffrage and admiration of the home-making virtues of women off-putting. Boyd argues that the emphasis on the maternal role found in these women could lead in two very different directions: it had helped to create a society based on male power with women related to the home, but the same talents could be applied in a 'wider sphere'. The complementary roles of men and women could be used in public life. The way in which Octavia Hill, Florence Nightingale and Josephine Butler exercised that role was radical and contradicted the seeming conservatism of their views.

Darley

The appearance of a new biography of Octavia Hill by Gillian Darley in 1990 reflects the continued interest in her work and provides an entertaining opportunity of comparing the tone of reviewers. Darley's biography was acknowledged to be meticulous and well researched and provided gentle criticism of Octavia Hill's views. The great majority of reviews were more sympathetic to Octavia Hill than in the 1980s, though often recognising her as a 'Thatcherite'. Several of the housing journals had reviews by women. Allen, in *Roof*, and McKerrow, in *Voluntary Housing*, both raised the possibility that Octavia Hill's reputation had suffered because she was a woman. 'If she had been a man would she have had such an inadequate and negative press to date?' (McKerrow, 1990). The male *Community Care* reviewer (Dossett-Davies, 1990) commented that she was 'no lefty', but he kept mainly to a description of the book. Ironically enough the most emotive phrase, 'a despotic do-gooder', came from the *Financial Times* (Chard, 1990) though the *Evening Standard's* 'Battle axe' ran it close (Porter, 1990). Hopefully the nature of these reviews reflects not only politics but also changing attitudes to women in the 1990s.

DISCUSSION AND CONCLUSIONS

The questions posed at the beginning of this chapter were: How extensive and substantial was Octavia Hill's influence on housing management? What exactly was the nature of that influence? Why does Octavia Hill seem to arouse so much controversy and strong feeling? Each of these will be considered in turn.

How do historians assess influence? How much of what is regarded as influential in the past is simply what historians have come to agree is influential? One source of evidence is what contemporaries regard as influential and by this standard Octavia Hill would be and has been awarded a place of some importance in the history of housing management. But there are plenty of examples of individuals given prominence by their contemporaries and down-rated by later generations. So the question of how historians make these reassessments still remains. Feminists have certainly argued that the work of women is consistently ignored and passed over wherever possible (see for example Bristol Women's Studies Group, 1979: 4, 5).

Another technique useful for assessing influence is tracing the process by which it was transmitted from one generation to the next. It can be argued that, in the case of the influences on housing management as a whole, this work has simply not yet been done. The lack of an adequate history of housing administration is one factor which has made the whole writing of this book more difficult. It is because of this lack of real evidence that two writers, critical of Octavia Hill and of similar theoretical viewpoints, can pose contradictory arguments about her. Spicker argues that Octavia Hill is a major and baneful influence on modern housing management while Malpass argues that housing management was recreated in the 1920s without the 'moralistic overtones' of Octavia Hill's method. Clearly both cannot be right. The present study traces the way in which the Society of Housing Managers encouraged the employment of women in housing and the dissemination of professional practice. It illustrates some of the ways in which 'Octavia Hill influences' moved into wider housing management. It also briefly discusses some of the other influences on the emergence of housing management, both inter-war and post-war. More detailed local work needs to be done but it is hoped that this present study will at least establish that to view Octavia Hill and her tradition as responsible for all that happened in housing management is as mistaken as the view that her influence was negligible.

The second issue was the nature of Octavia Hill's influence. There is a degree of consensus on some aspects of this. Both friends and foes acknowledge the value of Octavia Hill's emphasis on building up a relationship with the tenant, and on the importance of good management, getting repairs done. There is a lack of consensus on whether Octavia Hill is responsible for the rather authoritarian and dictatorial tendencies which appeared in many local authority housing departments and some associations. Octavia Hill was autocratic and could appear 'despotic' in her relationship with tenants. Yet the oldest interviewees mentioned, for example, that they were taught never to step over the tenant's threshold without the tenant's permission, i.e. not to invade the tenant's privacy, and that this came directly from Octavia Hill. This early direct evidence conflicts with many later views of the Octavia Hill tradition as encouraging incursions into the tenant's home.

So are we yet in a position to determine whether this practice came from the Octavia Hill tradition or, for example, from the 'public health' movement? Incursion into tenants' homes mainly occurred because of the need to check for 'bugs' and this was apparent to most authorities concerned with slum clearance, whether or not influenced by Octavia Hill (see Chapter 4). The evidence is not yet conclusive.

Similarly Spicker lays at Octavia Hill's door 'an exaggerated emphasis on rent payment . . . the use of notice and eviction as a primary sanction against tenants'. But the Society interviewees, including the oldest among them, laid great emphasis on the 'mutual obligation' view, i.e. good maintenance and management in return for rent, and the prevention of arrears rather than eviction. The consensus was very much that eviction was an admission of failure rather than a primary weapon and many contrasted this with modern local authorities' more frequent resort to eviction.

A conclusive decision on these issues could only be reached by careful and quantitative analysis comparing the practice of Octavia Hill influenced and non Octavia Hill influenced local authorities in the 1920s and 1930s. This has not been done and maybe never can be done because of the lack of sufficient representative evidence and methodological problems. It is not possible at this stage to get sufficient accurate and comparative material to come to definite conclusions. In this case, it is surely incumbent on commentators to review the available evidence but to be more open about the fact that their interpretation is going to be coloured by their own views.

The final question is, why does Octavia Hill arouse such strong feelings? Discussion here has to be speculative but the very emotive and irrational way in which the arguments have emerged indicates that the writers feel threatened or antagonistic with regard to some deeply rooted or underlying values. The two clusters of values most likely to be involved are political values or ambivalence about women's roles.

As far as political values are concerned, Octavia Hill's opposition to state intervention in housing is important. Most, if not all, of the writers in housing history have a left-wing perspective which often leads them to be very critical of people whom they see as holding up the 'progress' towards extensive state intervention in housing and subsidy. Leaving aside the question of viewing history as progress towards one specific point, it is logical to ask: are other people who hindered state intervention in housing attacked in the same kind of way? The answer must surely be no. For example, Macmillan is criticised for his policies in the 1950s but not subjected to the same kinds of personal attack.

Are the attacks on Octavia Hill more acrimonious because she was a woman? Angela Burdett Coutts, another female philanthropist of the period, with conservative views, is not attacked in the same way, but admittedly was less influential. The contrast with Florence Nightingale may be illuminating here. Modern nursing writers may be very critical of Florence Nightingale's ideas and some of her legacy but they acknowledge the contribution she made. Boyd argues that Florence Nightingale's life was used to reinforce a popular stereotype. To some extent her powerful personality and determination could be hidden behind the 'lady with the lamp' image. Octavia Hill's life could not be quite so easily adjusted to conform to the stereotype. Collecting rents and arranging for effective management and repairing of property does not have the 'ministering angel' image. Octavia Hill, like Florence Nightingale, suffered from inner conflicts, had bouts of invalidism and 'breakdown'. But she seems by later life to have resolved this far more satisfactorily both in maintaining human relations and in being able to delegate work and achieve out some kind of a balance between public and private life. Octavia Hill founded a movement which did have some influence and historical continuity – she was seen as successful and powerful. As a successful 'governing woman' she is open to the ambivalence which men feel about strong women without all the 'disguise' which is wrapped around Florence Nightingale.

For the purpose of this study, the way in which Octavia Hill has been seen both by contemporaries and by modern writers is of as much importance as the actual 'facts' of her life. The ambivalence towards the role of women in housing work is one of the key issues of this study. It is not possible quantitatively to assess the extent to which the tension between Octavia Hill's life and the stereotype of the woman's role contributes to the acidity of attacks upon her – and equally to the emotional response in support of her. It is hoped however that future discussion of Octavia Hill will at least take this possibility into account. The onus will be on writers to acknowledge bias, especially where lack of objective evidence or methodological problems makes it difficult to be conclusive.

3 The early years, 1912–32

MAJOR EVENTS IN HOUSING 1912–32

This chapter traces the gradual development of organised groups of women housing managers after the death of Octavia Hill in 1912. How did the 'Octavia Hill' influence spread? Why and how were women attracted to work in housing? What opposition was there and how was it countered?

In the nineteenth century, as a response to the appalling condition of working-class housing, a number of attempts had been made to help trusts and local authorities to build to improve conditions. The most effective of these was the Housing of the Working Classes Act of 1890 but 'of all new houses built between 1890 and 1914 less than 5 per cent were provided by local authorities' (Burnett, 1978: 181). It was only after the Housing and Town Planning Act of 1909 that local authorities were expected to become permanent landlords and it was the legislation of 1919 which decisively established their place in the provision of housing. Writers vary in the importance which they ascribe to the various factors influencing government attitudes to housing legislation: the effects of the First World War, rent control, social and political change, structural failure in the housing market and the political unpopularity of the private landlord (see, for example, Bowley, 1945: 3; Gauldie 1974: 296; Daunton, 1984: 708). The Conservatives were convinced that, at least temporarily, the private market could not cope and the state had to intervene, while the emerging Labour party had a more substantial commitment to public housing as a means for long-term improvement in housing conditions.

The Housing and Town Planning Act 1919 (the Addison Act) imposed on local authorities the duty of surveying the needs of their districts for houses and making and carrying out plans for provision of the houses needed. A subsidy was provided by the Treasury and the standards for houses were defined.

The framework for local authority housing was established, but the Conservatives viewed such state intervention as a temporary measure to cope with the effects of the war and social disruption. When they were in power they gradually shifted back to private market solutions and the Labour governments of the time were short-lived. By 1933 the local authority's role in housing was limited to that of slum clearance. Even the Conservatives accepted that slums would not be dealt with by the private market and by the 1930s the slum problems had become very pressing (Burnett, 1978: 237; Bowley, 1945). Despite the vicissitudes of political change, local authorities in this period had acquired a permanent housing stock. As we shall see, the switch to slum clearance had an effect on the types of household rehoused by local authorities and gave rise to calls for more professional housing management.

EARLY MOVES TOWARDS THE FORMATION OF AN ASSOCIATION OF WOMEN MANAGERS

> Miss Hill had, before her death in August 1912, so organised her work that each manager trained by her should be able to carry on independently when she was gone. Of the various estates which the Church (then the Ecclesiastical) Commissioners had put into Miss Hill's hands, one – 800 tenancies in Walworth – had always been in Miss Lumsden's charge. Other managers, Miss Covington in Westminster, Miss Mitchell in Southwark, and Miss Joan Sunderland in Lambeth, were in control so far as relations with tenants were concerned but went weekly to report to Miss Hill at her office. . . . At Miss Hill's death, these managers were put in touch with Messrs. Clutton and became directly responsible to them. . . . In the Notting Hill area . . . Miss Dicken, aided by Miss Perrin, showed a genius for managing 'difficult' tenants. A group of properties in Southwark, owned by Lady Selborne, was taken over by Miss Galton, who also took charge, later, of considerable estates in the Old Kent Road belonging to the Church Commissioners. . . . Miss Yorke . . . remained at 190 Marylebone Road and took over responsibility for most of the smaller outlying properties and centralised the work at this office where Miss Upcott worked for the first four years . . .
>
> (Upcott, 1962: 5)

This passage is quoted at length because many of these 'older managers' or 'those who had worked with Miss Hill' formed a key link in keeping women's work in housing management going during the period from Octavia Hill's death to the formation of the stronger, more formal occupational group in 1932. As some of their names reappear later, it is possible to trace the Octavia Hill influence gradually spreading outwards. The records for this period are more sparse and fragmented than for later ones. Not surprisingly, only two interviewees (Miss Upcott and Miss Larke) were involved in housing work in 1912, though a few others had begun to train or work by the late 1920s. There are a few other accounts of parts of the early work (for example, Brown, 1961) and these can be complemented by some primary sources, particularly the publications of the organisations concerned. However there are some periods, for example 1920–26, when such primary records are sparse. So the careers of individual workers make a valuable contribution to tracing the spread of influence.

According to Upcott (1962: 6), informal afternoon meetings held by Miss Yorke formed a way of keeping people together after Octavia Hill's death. Initially a meeting of these workers was held to discuss whether they should respond to the government's appeal to 'take a man's job' during the war. But the women felt that they could not take over the work of estate agents 'without changing the methods so as to conform with their own principles', and that when the men returned the management would return to former methods, 'as the women would be at least morally bound to hand back the Agencies' (Upcott, 1962: 6). So they decided not to take on these jobs but the meeting proved the spur to forming an association of their own. The Association of Women Housing Workers was formed in 1916. It comprised not only those who had recently been part of Octavia Hill's staff, but also women who had in the past collaborated with her. It had a council and a training scheme 'combining practical and theoretical training, with attendance at lectures, culminating in an examination for an assistant's certificate after a year's work and a manager's certificate after two more years of satisfactory work' (Upcott, 1962: 6).

The Association of Women Housing Workers' leaflet of 1916 stated that its aims were:

1 To unite all engaged in Housing Work.
2 To have a representative body to which all interested in Housing Work may apply.

3 To arrange for the training of workers and promote the advancement of the knowledge necessary for the efficient management
of house-property.

While the annual reports for 1916 and 1917 are printed leaflets, those
for 1918, 1920 and 1926 are in typewritten form only, possibly
indicating a period when the association was struggling for existence. Like many voluntary bodies, it initially operated on slender
resources. For example, 'Financially, the Secretary was able to report
a balance of £2.6.10d.' (AWHPM, 1917b). Although it may have had
a struggle to establish itself, this Society appears to have been the
most substantial grouping of Octavia Hill managers during this
period. By 1930 it was able to employ a secretary for at least four
days a week (AWHPM, 1930b). There was however another grouping
of women managers clustered around Miss Jeffery at the Crown
Estate Commissioner's Office at Cumberland Market, which became
the Octavia Hill Club, and another grouping, of municipal managers,
emerged later (Upcott, 1962: 7).

PROGRESS IN EMPLOYMENT

Work with the Crown Commissioners

Miss M. M. Jeffery, who had been Octavia Hill's secretary, in 1916
had been appointed to manage the Cumberland Market (London)
Estate of the Commissioners of Crown Lands (Later the Crown
Estate Commissioners). This was an estate 'of about 850 houses
divided into about 2,000 tenancies, occupied by a population of
about 7,000' (Parker Morris, 1931: 2). The Crown Estate Office
played a key part in publicising Octavia Hill's work and in training
staff. Miss Jeffery formed her own system of training (Anon., 1931).
There was obviously some degree of rivalry between Miss Jeffery and
the other group of managers. As one interviewee remarked, 'the
besetting sin of the two early societies [was that] they both thought
they alone had the truth, the direct line of apostolic succession'.
   Miss Jeffery was working on commission and had to pay staff out
of it. According to interviewees, she was evidently very good at
picking up and encouraging people who had some leaning towards
housing work, providing them with a starting point and encouraging
them to move into new posts in the provinces. Her club appears to
have had its own publication but only one copy of *The Octavia Hill
Club Quarterly* has been located so far. This is a duplicated booklet of

some substance, containing an interesting early account of the work
at Cumberland Market, St Pancras Housing Association and the LCC,
as well as a report of the conference of municipal managers (*Octavia
Hill Club*, 1928).

At first the estate was organised from an office in Lambeth but it
rapidly became necessary to form an office in one of the houses on
the Cumberland Market Estate.

> Then three rooms in No. 42 Cumberland Market were taken. Here
> indeed there was rather more space, but no shelves for the rapidly
> increasing number of files; and desks and chairs could not be got in
> quickly enough for the rapidly increasing work and staff. Collec-
> tors made up their money on the floor, and wrote their orders
> balancing their pads on their knees. . . . Looking back to those
> earlier days – to the rush and hurry of taking over houses in the
> intervals of collecting; of writing specifications on Sunday because
> there was no time during the week; of working often until ten
> o'clock at night – and then contrasting the present time of orderly
> organisation and arrangement, reveals a work of care and thought
> which has been well worthwhile.
>
> (Allen and Lawes Wilkinson, 1928: 4,5)

Work with the Ministry of Munitions

The Ministry of Munitions had built estates to provide permanent or
temporary accommodation for some of its workers, but had encountered problems of management:

> In the Welfare and Housing sections of the Ministry there were
> men acquainted with Miss Hill's successes in regenerating slum-
> property and, when some of the problems seemed beyond the scope
> of ordinary estate management, it was to her school of thought that
> they turned. Lord Dunluce, now Earl of Antrim, versed in housing
> knowledge through his work with The Peabody Trust, took the first
> step in suggesting the introduction of women and Mr. G. H. Duck-
> worth, already for a generation a supporter of the private side of the
> work, was, as Director of Housing in the Ministry of Munitions, a
> staunch and loyal friend to the women managers, not only during
> the war, but in the difficult and unpopular time afterwards.
>
> (Upcott, 1923: 32)

They approached the women managers' group and 'Miss Sunderland
and Miss Upcott visited the estate at Dudley, where they found rising

arrears and a low standard of maintenance' (Upcott, 1962: 6). A number of trained women housing managers were appointed in 1917. Miss Sunderland went to Barrow, Miss Lumsden to Dudley with Miss Upcott as her deputy. Subsequently, Miss Lumsden was given an advisory post at the Ministry itself while Miss Upcott took over the management of the Dudley estate.

Miss Upcott provides a good example of the background and work of one manager. She had followed a degree at Somerville with one year's training at the London School of Economics at a time when few of her contemporaries had heard of social work. She was attracted by Octavia Hill's reputation to become a trainee with her in the last year of her life, and carried on working with Miss Yorke after Octavia Hill's death. After a brief spell with another charitable organisation, she moved to Dudley. Miss Lumsden asked Miss Upcott to share the Ministry of Munitions work in Dudley and as Miss Lumsden took over more national duties Miss Upcott ran the estate in Dudley.

Miss Upcott's account of her work in Dudley gives an idea of the energy and innovation which she brought to the job and of her willingness to consider the wider aspects of estate life. In her interview she said that she spent three and a half years at Dudley.

> 'when I took over Dudley it had been managed by somebody who really didn't know anything about it and about half the tenants were in arrears so we had to start collecting rent. When I left there were . . . about four pounds in arrears. We were in entire charge of the housing estate. We did the letting and we did the accounts, reported to the Ministry, collected the rents.'
>
> (Upcott, 1979)

Dudley was a temporary estate. 'It was built of either rather flimsy concrete cottages or else wooden and there were no paths or railings or anything, everything had to be tidied up.' Miss Upcott had to take on an assistant and they were concerned about developing community facilities on the estate as well as the physical aspects. 'We didn't know any other line so we started a Women's Institute.' This flourished and then Miss Upcott took steps to try to obtain a district nurse for the estate, since the organisation in Dudley which provided such nursing felt they could not take on the estate. The nurse's salary, £130 a year, had to be guaranteed. Failing to obtain help from more wealthy people in Dudley, Miss Upcott took on the responsibility herself 'and the tenants were awfully good, they rallied round and they had whist drives and sales of work, bazaars and all kinds of things you see and we raised that £130 and we got a nurse for ourselves, that was a great thing because at that time there were practically no social organisations you see, so we had this organisation which was very successful and they promoted Boy Scouts and Brownies and one tenant took over the Boy Scouts because she was interested in this. We had a certain amount of boxing. I was a proprietor of a boxing club which shocked a good many people.'

(ibid.)

Sir George Duckworth instituted periodic meetings of the managers employed by the Ministry.

The value of the work was shown by the great improvement in standards and the avoidance of rent strikes which were prevalent at the end of the war. The Ministry seemed satisfied and Miss Lumsden devised a training scheme to ensure continuance of this type of management.

(Upcott, 1962: 7)

However, this attracted the notice of the magazine *John Bull*, which was not in favour of female workers:

The Housing Department of the Ministry of Munitions . . . is having the impudence to train twenty women . . . as house and estate agents. Of course the Ministry has no houses to speak of, but it is training these females to manage working-class houses. They are provided with a course of lectures at Battersea Polytechnic. . . . When qualified the women will become Superintendents, at £250 a year. This cold-blooded, premeditated throwing-away of public money is the most outrageous so far. The whole project is so absolutely absurd, futile and unnecessary. First of all, if such jobs are to be created, men should have them.

(*John Bull* Editor, 1919c)

The controversy about this issue even led to a question in Parliament (Ministry of Health, 1919c). The attack in the press evidently squashed any formal plans for continuing this type of management. Miss Upcott said:

'well of course we didn't like it but we felt we couldn't stand out against the returned soldier . . . they had some claim to jobs and we should be very unpopular if we'd been thought to be obstructive to their finding jobs and we felt that we couldn't really stand out about that.'

(Upcott, 1979)

But, though this experiment came to an end, it was subsequently seen as having been important. It had demonstrated that women could cope with larger-scale management for a public authority and could 'pull up' estates that had begun to deteriorate. It is probably no accident that two women housing managers to be appointed to local authorities in the 1920s (Miss Upcott and Miss Geldard) had both worked for the Ministry of Munitions.

Contact with central government and the Ministry of Health

Even before the work with the Ministry of Munitions got under way fully, the women housing managers had been trying to further more general contacts with central government. The Association of Women House Property Managers' Annual Report for 1917 says:

> We have also had an interview with Sir M. Bonham Carter at the Ministry of Reconstruction. It seems possible that any rebuilding schemes after the war will be put into the hands of the local authorities with a Government subsidy and a central body behind them. . . . We have also got into touch with the Housing Committee of the Local Government Board . . . and Mrs. Rawlins and Miss Jeffery have appeared before the Committee to discuss plans of working class houses.
>
> (AWHPM, 1918b)

In 1918 'Miss Jeffery is on Lady Emmott's Committee of the Reconstruction Committee' (AWHPM, 1919b). This was the Women's Housing Sub-Committee which had been set up at the Ministry of Reconstruction following pressure from women's groups, of which AWHPM was one (McFarlane, 1984: 27). The Ministry had been trying to encourage women's participation in matters to do with housing. Following on from the work of the Women's Housing Sub-Committee, Circular 40 in 1919 advised local authorities 'to take such measures as are practicable to obtain the views of women' (on matters to do with provision of amenities and house plans). It advocated the co-option of women onto housing committees and the setting up of women's advisory committees (Ministry of Health, 1919e). The Minister of Health, Christopher Addison, took a deep personal interest in housing, and the Ministry had begun to issue a housing journal from July 1919. Women managers and the desirability of their work are mentioned in a number of issues of this journal as is the desirability of women's advisory committees at local authority level (for example, Ministry of Health 1919a, Ministry of Health, 1920a and a number of others).

In July 1920 *Housing Journal* devoted its major article to 'The management of property,' pointing out the need both for good management and for training for the managers:

> Little is done except by the Association of Women House Property Managers, who have rendered such admirable service in redeeming unfit property.. . . Property management is a profession as well adapted to women as to men.
>
> (Ministry of Health, 1920b: 1)

A further article in the same issue on 'Property management as a solution to the slum problem' describes the work of 'Octavia Hill' managers in some detail and mentions that Neville Chamberlain's Committee on Unhealthy Areas had suggested that local authorities purchase unhealthy areas on a large scale and entrust them to the care of women managers (Ministry of Health, 1920c: 6).

The second report of the Unhealthy Areas Committee, which was published in 1921, also argued the case for improved management. 'Considerations of this kind draw us to the conclusion that the management of old property on the Octavia Hill system . . . might be extended with advantage to the community' (Ministry of Health, 1921). This advocacy of women managers by the Ministry of Health displeased the Secretary of the Surveyors' Institution, who felt that the articles

> overlooked the fact that there is such a body as the Surveyors' Institution, the members of which manage a large proportion of the house property in Great Britain and, I believe, do so to the advantage both of owners and tenants. The women house property managers referred to in the article have done good work . . . but I think it unfortunate that the writer should have appeared to indicate that they alone were moving forward in this matter.
>
> (Goddard, 1920)

After 1921, the journal *Housing* was reduced to a monthly issue and then stopped, parallel to the cuts then being imposed on the housing programme. AWHPM continued to be invited to give evidence to official committees – for example, their 1920 report mentions giving evidence before the Committee of the Ministry of Health 'on the operation of the Rents Restriction Act'. Contacts clearly were maintained; for example, in 1926 'Mr. George Duckworth, CB, formerly Controller of the Munitions Housing Schemes', was elected a vice-president of AWHPM (AWHPM, 1927).

Women's employment in local authorities

Two members of the Association went to Birmingham at the end of the war. In 1921 Amersham Rural District Council had 'appointed Miss Geldard as the best candidate from a list of applicants of both sexes, but with no actual adoption of the principles derived from Octavia Hill' (Upcott, 1962: 6). In 1923 Upcott commented: 'The Amersham Rural District Council, beginning with a few scattered rural cottages, is testing the value of a woman manager, and such a development seems to offer great promise for the future' (Upcott, 1923: 34)

For a time there was not much further progress until the Association took a more active role in encouraging the employment of women. One interviewee commented that in the late 1920s

> we circularised a great many authorities and nobody took any notice except Sir Parker Morris (who wasn't Sir then), and he had the brilliant idea of bringing up his entire committee to inspect our work. The committee were taken to Miss Jeffery's estate at Cumberland Market and to the Church Commissioners' estate of Walworth. They were very impressed at what was being achieved with old and run-down properties.

The committee decided to appoint an Octavia Hill manager of a particularly difficult estate, and Miss Upcott left.

In 1931 Parker Morris described some of the factors which contributed to the decision to employ a woman manager:

> At Chesterfield the Corporation had carried out (after the war) a slum clearance scheme and removed the tenants to a new housing estate. After a time it was found that the condition of many of the houses was deteriorating and the trees, fences, and grass margins were suffering serious damage. Disturbances amongst the tenants were not infrequent. Heavy arrears of rent accumulated during the coal dispute in 1926. After one or two instructive experiments, the Corporation decided to investigate the Octavia Hill system, and, as a result, they enthusiastically adopted the system for their estate to which the tenants from the slum area had been removed. After the

system had been in operation for two years it was found that 90 per cent of the original tenants had shown definite and more or less continuous improvement in the condition of their homes.

(Parker Morris, 1931: 3)

At Chesterfield Miss Upcott managed an estate which housed tenants who had come from slum clearance and were rather poorer; 'there was a man there also who did the more prosperous lot'. When Miss Upcott left she was replaced by another member of the Association (Upcott, 1979).

Encouraged by the example of Chesterfield, similar appointments were made at Walsall, Chester, West Bromwich, Stockton-on-Tees and Rotherham. Parker Morris seems to have been influential in advocating this change and he remained a staunch friend of the women housing managers throughout the history of their organisation. Another example of the influence of the older offices can be seen in Kensington, where the Medical Officer of Health had been so much impressed by the results achieved by Miss Dicken at the Improved Tenements Association that she was asked to manage some borough property in Notting Hill prior to the appointment of an Octavia Hill manager by the Council (Galton, 1959).

Private employment and trusts

It is important to remember that the early housing managers working for private individuals and even for the Church Commissioners were working on commission (calculated on rent collected) and paid their staff out of it. Kathleen Brown, commenting later on the superior conditions at Kensington Housing Trust where the committee had invested in a pleasant office, said:

In the early days many an Octavia Hill trained manager, paid on commission, had to find the office furniture and equipment out of her own pocket. The office would be the front parlour of a tenement house and the student would work at the schoolroom table from the manager's old home and the telephone, when this was installed, would stand on a little occasional table from some forgotten drawing room.

(Brown, 1961: 6)

The scope of the work also remained limited in some instances. For example, with the Church Commissioners, 'all capital questions, general level of rents, financial policy generally, are kept in the Commissioners' own hands' (Upcott, 1923: 22).

Nevertheless, work with the housing associations and trusts continued to be important for the women managers, especially as new societies were formed in the 1920s and 1930s as a response to the appalling housing conditions (Allen, 1981: 48). Those linked with Octavia Hill work included the St Pancras Housing Association and the Kensington Housing Trust in London, and the Liverpool Improved Homes and Birmingham Copec (Conference on Christian Politics, Economics and Citizenship) in the provinces (Barclay, 1976; Brown, 1961; Fenter, 1961; Baskett, 1962).

Annie Hankinson wrote about the work of the Manchester Housing Company Ltd both in an article and in a letter to *Housing* in 1919 (Hankinson, 1918, 1919). This society had been formed to buy and manage property on Octavia Hill lines and was also prepared to manage for other owners. By 1919 they had 150 properties and were operating along financially sound lines.

The Birmingham Copec House Improvement Society was formed after a national conference of Christian churches there in 1924 and as a response to the very poor housing conditions in Birmingham. Women such as Anne Smith, a factory inspector and Florence Barrow, a Quaker, were important in getting a group interested in housing action established. Some of the women, including Miss Lidderdale, who had been trained under Octavia Hill, initially collected rents on a voluntary basis, but as the number of houses grew it was recognised that a full-time paid worker was needed. Among others, Miss Jeffery was consulted and in 1926 Miss E. M. Fenter, a graduate who had trained with Miss Jeffery, was appointed. (She continued with the Society for many years until retirement in 1953.) (Fenter, 1960: Chapters 1 and 2).

The Women's Pioneer Housing Ltd, formed in 1921, was a co-operative society with a more specific focus on providing housing for women who needed to make their own homes. Its scale of operation remained small (Women's Pioneer Housing Ltd, 1935).

Liverpool Improved Houses Ltd was formed in 1927 to help the poorest tenants. They invited 'Octavia Hill management' from the start and also took on management for private owners, though by 1928 the number of houses managed was still small. At this stage there were still 18 women listed as members compared with nine men (Liverpool Improved Houses Ltd, 1928). Eleanor Rathbone was involved in launching the Association (Allen, 1981).

The St Pancras Housing Association had been formed in 1924,

largely through the work of Fr. Basil Jellicoe. Irene Barclay, who had been trained by Miss Jeffery, was appointed as its secretary in 1925 and this Association went on to make a substantial contribution to housing in its area (Barclay, 1976: 17).

Women were playing quite a prominent part both in forming and running these new housing associations, but little attention has been paid to this contribution. It is acknowledged in Patrick Allen's unpublished account of the development of housing associations (Allen, 1981) but little detail is given.

THE WORK OF THE WOMEN HOUSING MANAGERS' ASSOCIATION

Organisation of the Association

The Association of Women Housing Workers started off with a council which consisted of 25 members, a committee including chairman, secretary and seven other members. Since the total membership listed was only 50 this seems a rather top-heavy kind of organisation (Association of Women Housing Workers, 1916a). By 1926 the Association had a president (Viscountess Astor) and three vice-presidents: the Countess of Selbourne, The Lady Emmott and Miss Rosamund Smith (AWHPM, 1927). Three more vice-presidents, Sir Stanford Downing, Secretary to the Ecclesiastical Commissioners, Captain Townroe, Housing Correspondent of *The Times*, and Mr George Duckworth, CB, formerly Controller of the Munitions Housing schemes, were elected in that year (AWHPM, 1927).

By the 1930s the Association had a full-time secretary, a Council, Executive Committee and two sub-committees. Junior members' meetings had also been held 'at the sanction of the Executive Committee' (AWHPM, 1931b). Since minute books are not extant it is not possible to see in any more detail how this organisation actually worked. The appointment of presidents and vice-presidents was clearly used as part of the Association's propaganda activities, keeping up contacts with influential people. Lady Emmott, for example, was a contact with the National Council of Women and chair of the Housing Sub-Committee of the Ministry of Reconstruction in 1918 (McFarlane, 1984: 27).

Training

The formal schemes

In its first leaflet in 1916, the Association of Women Housing Workers had given training as one of its aims and this emphasis was continued by AWHPM. By 1917 AWHPM had been able to draw up a training scheme and had two workers in training. 'The scheme provides for those taking up the work as a profession and for those wishing an insight into Housing Work as part of other social training' (AWHPM, 1917b). A year later the Association had decided that a fee should be charged for training. However, the report for 1918 noted a difficulty which was to continue for some years: 'We offer a long training with no pay and no definite prospect of work at the end of it, not encouraging conditions for anyone having to earn their own living' (AWHPM, 1919b).

By 1920 there was an active Training Committee with a training scheme and a syllabus for an examination. Those who passed it were awarded Assistants Certificates' (AWHPM, 1920a). By 1926 there were six students in training, two of whom passed the Intermediate Examination of the Surveyors' Institute (AWHPM, 1927). The printed training scheme for AWHPM which is extant appears to belong to this period. This mentions the AWHPM scheme for theoretical and practical training and examinations for Assistants' and Managers' Certificates after one and two years' training respectively.

> Girls who have matriculated should consider taking the BSc London degree in Estate Management, coming to the Association for practical insight into the work both before and during their college course.
>
> (AWHPM, 1921–1926)

By 1929 there were 11 students whose names are given in the report (AWHPM, 1930b) and a number of them went on to become important in the later work of the Society and housing: for example Helen Alford, who became housing manager at the Royal Borough of Kensington (Alford, 1981), and Margaret Hurst, who became the first woman housing manager in South Africa in 1934 and subsequently the first housing management adviser to the Ministry of Health in 1943 (Hill, 1981). The training scheme had been revised and reprinted. This scheme provided for candidates being interviewed by the Training Committee. The fee to the Association was 20 guineas and, in addition, students would have to meet examination and lecture fees. Training was for two years and included practical and theoretical work. Students were required to qualify for either the Woman House Property Managers examination of the Chartered Surveyors' Institution or the BSc Estate Management, University of London or the Sanitary Science Certificate of the Royal Sanitary Institute. The syllabus included Building Construction, Estate Accounts and Law of Landlord and Tenant as well as requiring approved training and practical experience. This method of examination was to become the major means of theoretical training in the women housing managers' societies.

The experience of interviewees

In the very early days, practical training was on the job but theoretical training was more difficult to come by. Larke, whose family 'built and owned houses' and who 'was always more at home with hammer and chisel than with needle and thimble' (Larke, 1981) recalls that prior to 1915

> 'Somehow I heard of a class for Housing Management taking law, technical work, sanitation and so forth. It was being taken by an eminent surveyor and sanitarian while he was studying for his barrister's degree. Very soon during the late afternoons I found myself on a high stool among 20 young men listening to lectures delivered in a very dry way, but this man, W. T. Cresswell, was interested to know why I was there. He soon suggested I sit for the exam of the Royal Sanitary Institute and set me to studying sanitary appliances at the nearby Sanitary Institute. I enjoyed reading the law subjects most and after a second try obtained my M.R.San I. One other woman had this.'

Another interviewee remembered:

> 'you either became a surveyor or did BSc Estate Management or you could take a ''sanitary inspector's thing''. The surveyor's was a long-term thing – didn't want it, really. So I took this other horrid thing – really meant for clerks of works and builders – so we went round looking at sludge, drains and roofs. Got hold of a builder and got taken round.'

The surveyors' examinations had only just become open to women; Miss Jeffery encouraged Irene Barclay to sit for them and Misses Barclay and Perry became the first women surveyors following the Sex Disqualification (Removal) Act of 1919 (Barclay, 1981).

By 1930 the Association had acquired a permanent office but was still working on a very modest basis. Nevertheless such an office was a help in recruiting some of the educated women who were beginning to emerge from the universities and looking for information about employment. The Association had also begun to arrange for people to visit offices to help them to decide whether housing management was a suitable career. This practice was to continue and was particularly helpful to those who knew little about housing.

Another interviewee, Miss Lamplough, had come into housing work after taking the BSc in Estate Management; she was among the first 30, and the first woman, to get that qualification. At that stage it was not easy for a woman to find work in a private surveyor's office and Miss Jeffery persuaded her to try housing management (Lamplough, 1981). For most people, however, the full surveying qualification was too much to contemplate and, until a properly relevant examination was established, theoretical training was probably rather haphazard.

Although great stress was laid on practical training on the job, especially in the early stages, this could be less than satisfactory. One student, who was in five offices in one year, felt that she could not learn much moving round so fast.

Even in Miss Jeffery's office the training could be rather haphazard. One interviewee said:

'Now it was curious training. It was very good on the personal side because that depended very much on who you were taken round with. It was a very interesting estate, it had some ghastly property which was going to be pulled down in time . . . on the repair side because you had these small contractors, caretakers and so on and all that was extremely good experience . . . but how much people learnt on repairs depended on who they got taken round with.'

Publicity

The women housing managers from an early stage realised that publicity was a key issue. AWHPM's first report states: 'The difficulty is still one of getting in touch with more property owners' (AWHPM, 1917b). By the next year decisions were being taken about circulating a standard leaflet more widely, probably *Working-class houses under ladies' management* (Association of Women Housing Workers, 1916b). Copies were sent to organisations inside

London and to national bodies. Pamphlets and notices of training were sent to colleges and settlements (AWHPM, 1918b).

From an early stage small leaflets about the work seemed to be used as a major means of publicity but, as these are undated, it is not possible to be entirely confident about the sequence and it is evident that material was re-used (Yorke and Lumsden, undated; Association of Women Housing Workers, 1916b). It seems likely that a basic leaflet of this type was used throughout this period, supplemented by other leaflets written by members which were republished or circulated by the Association.

Opportunities were also taken to provide articles for newspapers or journals or to write letters to them. For example, as early as 1913, Annie Hankinson had gained local publicity for the work of the Manchester Housing Company (Hankinson, 1913) and her later article for the *Woman Citizen* was subsequently republished by AWHPM (Hankinson, 1918). An article in *The Times* by Sir Reginald Rowe, entitled 'A work of slum reclamation', was an account of the work of the Improved Tenements Association but also discussed the work of women managers extensively (Rose, 1931). In February 1931 *The Times* published 'Women as estate managers', discussing their work with approval and mentioning the training given by the Association and by Miss Jeffery (Anon., 1931).

Miss Jeffery produced a number of publications, and a *Memorandum upon property management and slum clearance* by Parker Morris, then town clerk of Westminster (Parker Morris, 1931) was circulated by the Society. This pamphlet was specifically directed towards local authorities' slum clearance role.

Extensive use was made of opportunities to present papers to other organisations. Papers presented by Miss Geldard to the Rural Housing Association, by Miss Thrupp to the Royal Sanitary Institute (Thrupp, 1929) and by Jean Thompson to the Institute of Public Administration (J. Thompson, 1931) became leaflets which could be used for publicity.

Links with other organisations

The first extensive list of contacts with other organisations is found in the Association of Women House Property Managers' Annual Report for 1930. This mentions seven meetings which had been addressed by members over the year, covering a fairly wide range: the Royal Sanitary Institute Congress, Wrexham Borough Council, Women Citizens' Association, women's clubs, local authorities, and a branch of the Auctioneers' and Estate Agents' Institute. Mention is also made of a number of conferences and meetings attended by the secretary and other members (AWHPM, 1931b).

The Association seems to have had strong links with some women's organisations. We have already noted that there were women's groups with an interest in housing and that the women housing managers had joined with them on the Women's Housing Sub-Committee of the Ministry of Reconstruction. It was clearly important to keep in touch with this constituency. It is likely that many of these contacts were informal; for example there are early mentions of correspondence with womens institutes and the Association was itself situated for a time at the Women's Institute at Victoria Street (AWHPM, 1918b). In 1926 the Association sponsored a motion at the conference of the National Council of Women about the reconditioning of older houses (AWHPM, 1927) and participation of this kind continued through the period.

Such publicity and links were important not only for contacting employers but also for attracting recruits at a time when housing management was little known. For at least two interviewees the Women's Employment Federation provided a useful source of referral to AWHPM after they had been attracted to housing by newspaper articles. For others, informal contacts via friends and family or meetings provided the way in. A tutor at Somerville was instrumental in getting two of the interviewees interested in housing.

THE MOVE TOWARDS ONE ASSOCIATION

In 1928 Miss Upcott held a successful conference at Chester which was attended by practically all the municipal managers and their assistants. 'But this conference emphasised the fact that a regrettable division had come about in the professional grouping of Octavia Hill trained women' (Upcott, 1962: 7). There was a strong feeling that one association should be formed and discussions about this started around 1930. Even at this stage, however, there were difficulties. The intense personal influence of Octavia Hill was still strong; discussion centred around questions of who best represented her tradition:

> The question of qualifications for membership had engendered some controversy, some wishing to limit this to those actually engaged in paid professional work, others remembering the old

voluntary workers who had been the backbone of Miss Hill's early work. Finally, it was agreed that anyone who had done bona fide housing work under Octavia Hill should be equally eligible with the present professional managers.

(Upcott, 1962: 7)

The disputes over principle seem to have been exacerbated by a degree of personal rivalry. To the younger managers and to those out in the provinces, who felt a great need of support, the need for one association was so obvious that they felt impatient with these discussions. 'I can remember thinking that it was about time that they stopped arguing . . . over things which one or other societies had . . . things they were hanging on to . . . and forget all this back history' (Interviewee). Agreement to the form of amalgamation was finally achieved; the Society of Women Housing Estate Managers was brought into being and the assets of the previous societies were handed over to it (Upcott, 1933).

THE ACHIEVEMENTS 1912–32

At the time when Octavia Hill died all that remained was a scattered group of women who had been trained by her. By the end of the 1920s they had already gathered into groups and provided support for members to penetrate new areas of employment. Employment at the Crown Commissioners and at the Ministry of Munitions showed that women could handle housing management for large public bodies. From an initial situation where local authorities did not employ women in housing management, posts had been obtained in 14 authorities (though this remained only a tiny percentage of housing authorities). Employment in the housing associations was also increasing.

Parker Morris' *Memorandum*, published in 1931 by the National Housing and Town Planning Council, provides a useful summary of some of the gains in employment. After mentioning the Church Commissioners and Crown Commissioners he says that Rotherham, Chester, Cheltenham, Kensington, Leeds, St Pancras, Stockton, Walsall, West Bromwich, Bebington and Bromborough employ trained women property managers, and also mentions new employment at Westminster. The 1932 Annual Report of the Association notes appointments at Cambridge Housing Society, the Borough of Cambridge, United Women's Homes Ltd, Tunbridge Wells, Westminster Housing Trust, Bethnal Green and East London Housing

Association and that the Councils of Brighton and Newcastle under Lyme were also advertising for women estate managers.

Both the Association of Women House Property Managers and the Octavia Hill Club had managed to use the press to publicise the work of women managers and the Association had produced a number of its own publications. At the beginning of the period efforts had been made to set up a training scheme but these were hampered by lack of an examination and theoretical instruction. By 1930 more appropriate examinations had been arranged and the training system had become more formalised.

There was still opposition to this expansion of women's employment. The Annual Report for 1932 states:

> The Fulham Borough Council took evidence upon Octavia Hill methods and interviewed one trained woman, but a man superintendent was finally appointed. . . . The Housing Committee of the City of Liverpool recommended the appointment of a Woman Housing Inspector with two assistants but the matter was finally rejected by the City Council.

The argument that women had a 'special aptitude' for this kind of work was being used, but this could lead to them being diverted only to the management of the 'worst properties'. It would be some time before there was agreement with the Ministry of Health's statement that 'Property management is a profession as well adapted to women as to men' (Ministry of Health, 1920b: 1). However, a sound basis both in internal organisation and in employment had been provided on which the new Society could build.

4 1932–38: The employment of women managers

THE HOUSING BACKGROUND

The period 1932–1938 was a crucial one for the development of housing work and for the Society of Housing Managers. Major developments in housing policy provided opportunities for women and they began to be employed in greater numbers but there was still much opposition to face.

Under the 1920s' 'general needs' legislation many (though not all) authorities had been very selective about their tenants. But in the 1930s the emphasis on slum clearance was to make this more difficult. A memorandum by Parker Morris published by the National Housing Planning Council in 1931 pointed out this change:

> The tenants of the houses to be erected to replace the slums will be those residing in the slums. Many, in fact a large majority, of these families have a fairly high sense of self-respect and house pride, but a proportion (maybe 10 per cent) have little or none. Some of them are irregular in their employment and notoriously thriftless: they are apt to get behind with their rents and, if given the chance, to pile up hopeless arrears . . . The influence of such a comparatively small number of dirty, destructive, rowdy and thriftless tenants is out of all proportion to their number. They not only allow their own houses to become dirty and untidy, but they may cause serious damage to the fences, trees and grass margins. They are apt to cause rowdyism on the estate and to provoke dissension generally amongst their neighbours. As a result, the estate may get a bad name which may take years to outlive.
>
> (Parker Morris, 1931: 2)

Parker Morris went on to argue for the employment of 'Octavia Hill trained Women' to deal with these problems. We have seen in the previous chapter how women housing managers had been employed in cases where local authorities had felt there were 'problems' arising from their estates, and some of these were slum clearance estates. It seems reasonable to hypothesise that the swing to slum clearance in the 1930s may have provided further opportunities for the employment of women managers. Detailed local studies by Ryder (1984) and Dresser (1984) have shown that this situation could vary locally but confirm the increased need for attention to the rehousing of slum dwellers in the 1930s. It was certainly a spur to growing attention to housing management.

The Moyne Committee

The Conservatives nevertheless still looked for a time when government could withdraw from subsidising public housing. A departmental committee on housing was set up in 1933 under the chairmanship of Lord Moyne to consider arrangements for building and managing public housing. The Society contributed evidence to the Moyne Committee and tried to make their work known.

The Moyne Committee report strongly supported the establishment of a National Housing Board and suggested the setting up of local house management commissions. Both of these proposals were strongly attacked by representatives of the municipalities. On the other hand *Municipal Journal* commented favourably on the report's attention to management:

> In the future history of the housing developments of the 20th century Lord Moyne will receive an honoured place, because he and his Committee have at last emphasised the importance of management in dealing with working-class housing which can be reconditioned. The keynote of the whole report is trained management.
>
> (*Municipal Journal*, 11.8.33)

SWHEM Quarterly Bulletin quoted some of the recommendations on management in full:

> *Employment of House Property Managers*. We attach great importance to the employment, wherever practicable, of properly trained house property managers. We have no wish to exclude from this service the employment of trained men, but management on the Octavia Hill system involves assistance to the housewife in a number of different ways, and we think that it is therefore

desirable that women housing estate managers should usually be employed. The right kind of woman manager has a special aptitude for this class of work.

(*SWHEM Quarterly Bulletin*, October 1933c)

Moyne's proposals for a national housing commission were rejected by the government in favour of an advisory committee.

The Government have also decided to set up a strong Advisory Committee. All connected with local government will await with interest an announcement of the powers to be given to this Committee, and how far it will differ from the former Housing Council, established by Dr Addison in 1919, which passed away in 1921, unheard of and unsung.

(Townroe, 1934)

Although the proposal was described as a 'strong advisory committee', the Central Housing Advisory Committee, which was set up by the 1935 Housing Act, had very limited powers. Once again matters of housing management were left to the local authorities, with some exhortation from central government.

The Central Housing Advisory Committee and women's representation

The CHAC was set up in November 1935 under the chairmanship of Lord Balfour of Burleigh. Its membership was 25 including three women (Mrs M. M. Cooler, JP, 'Labour representative'; Miss Megan Lloyd George, 'MP for Anglesey'; and the Countess of Limerick, 'formerly Chairman, Public Health Committee, Kensington Borough Council': Ministry of Health, 1935a). However, some women did not think this representation sufficient. In December 1935 a letter was sent from Lady Sanderson stating that

A representative Housing Conference of Women's Societies (the third of its kind) has recently been held and I, as Chairman, have been instructed to write to you to ask if you will kindly receive a Deputation.

(Sanderson, 1935)

The Society of Women Housing Estate Managers was formally represented on this Housing Conference of Women's Societies. Also, at least seven of the housing associations represented had

Society members. The suggestion of representation was received without enthusiasm at the Ministry by the civil servant concerned:

> I presume that the Minister had better be advised to see this deputation as that course is likely to cause less trouble in the long run. Lady Sanderson was a member of the Women's Housing Sub-Committee of the Ministry of Reconstruction Advisory Council . . . and her suggestion of what the proposed Women's Sub-Committee should do seems to be largely taken from the matters dealt with in the Report of the previous women's subcommittee. . . . I think there is no real justification whatever for suggesting that there is a special woman's point of view in relation to such planning matters as the internal fittings of a house etc.
>
> (Ministry of Health, 1935b)

His correspondent wrote back in agreement:

> There was also a Women's Sub-Committee of the short lived Advisory Council under the Housing Act. . . . I was Secretary of it and see no reason for repeating the experiment.
>
> (ibid.)

The Minister seems to have been of the same mind. Though the deputation took place, Mrs Barclay being one of the representatives, the Minister thanked them gratefully but gave a bland answer, suggesting that the Conference should themselves set up a small sub-committee and he would be happy to arrange for co-operation and consultation between this sub-committee and a sub-committee of the Advisory Committee (Ministry of Health, 1936).

Some evidence from the Women's Advisory Housing Council, which was set up as a result of this consultation, was considered by the Balfour Committee, for example on playgrounds and gardens (Ministry of Health, 1938). It is worth quoting the Balfour Report's response to one of the items put forward by the committee.

> The Women's Advisory Council stated that working women were prejudiced against living in flats, their prejudice being due in part to the absence of proper places for the children to play; and that it would, moreover, be helpful if rest gardens were provided for the older people and, if space allowed, gardens which the tenants themselves could cultivate. . . . We should suppose that no local authority would wish to withhold from tenants amenities of the kind indicated, but unfortunately it is a problem of economics.

Land in the centre of London and in the big cities is scarce and commands a very high price. Flat construction is therefore expensive. As every additional amenity must inevitably be reflected in additions to the rents, the cost of any scheme for building flats must be a limiting factor in the provision of extra amenities. A good deal can be accomplished by means of window boxes in improving the appearance of both the home and the street; and as the cost is not great, we recommend that local authorities be asked to give consideration to this matter.

(CHAC, 1938: 34, 35)

In this case the women might seem to have had better arguments on their side than the rather condescending experts on the committee.

The existence of the Women's Housing Conference and formation of the Women's Advisory Housing Council indicates that there was some 'community' of women's interest in housing. The history of the 1918 Women's Housing Sub-Committee has been traced by McFarlane (1984). Despite this continuing effort women's views remained under-represented.

The Balfour Committee

The Balfour Committee (a sub-committee of CHAC) was appointed in January 1936

to consider, inter alia, the general question of the management of housing estates by local authorities with special reference to the employment of trained house managers.

(CHAC, 1938: 5)

Lord Balfour of Burleigh was chairman of the committee. He was also chairman of Kensington Housing Trust from its foundation to 1949 and thus a friend both of women housing managers and of housing associations.

In April 1935 the committee heard that

The Minister had recently received a deputation from the Institute of Housing Administration which was in a sense a rival body to the Society of Women Housing Estate Managers, though the methods and management advocated by the two bodies differed in certain important respects. The Institute has asked to be directly represented on the Advisory Committee.

(Balfour Sub-committee, 1935)

The sub-committee decided to ask both bodies to give evidence but SWHEM, as 'the older and better known body', was to be asked to appear first (ibid.). Both bodies gave evidence and submitted written memoranda. Evidence from Sutton Dwellings Trust, The Peabody Donation Fund, London County Council and the AMA was also considered. A questionnaire and visits were used to get information. Lord Balfour felt that 'If the enquiry was to produce reliable results, continuity in the examination of houses was of first importance' (Balfour Sub-committee, 1937). The committee carried out the most thorough enquiry into housing management that was to occur for many years. Tenants' views, however, were little represented. There was a report from the chairman and members of Hulme Tenants' Advisory Committee, which consisted of 'persons interested in social matters in the Hulme area' and had been formed to 'act as a link between the Corporation and the tenants of the Hulme Clearance Areas' (Hulme Tenants' Advisory Committee, 1937). But this was exceptional. Nevertheless it is clear that the evidence from the visits convinced the committee that housing management in some local authorities was in need of improvement.

The Society of Housing Managers' evidence gave prominence to the idea of housing management as an integrated function with the services affecting tenants, and visiting them, being carried out by one trained person – and emphasised the suitability of women for this work (Balfour Sub-committee, 1936; Thompson, 1936).

The Institute of Housing identified different types of housing officials with different ranges of duties. It also argued for training for the work and outlined the Institute's new examination scheme. Under the heading of 'welfare work', to assist the housing manager in his duties it advocated the appointment of a woman welfare worker.

> It is considered advantageous that she lives on an estate and must possess very definite qualifications, be of the motherly or matronly type with a definite love of welfare work . . . She is, therefore, always available to render all the assistance that lies in her power by closely co-operating with the Doctor, District Nurse . . . and other services.
>
> (Institute of Housing, 1936a)

The stereotype of women's role can be seen in full force in these arguments.

Evidence from the Peabody Trust described their 'resident superintendent' model of housing management based on the use of ex-servicemen (Agate, 1936). Evidence from local authorities showed a variety of arrangements, ranging from large bureaucratic systems such as Manchester's, through systems in which control was split between different departments, such as in Cardiff, to the use of agents for housing management. This latter practice did not however meet with the approval of the committee or of the Ministry of Health (Howes, 1937).

The Balfour Report and its effects

The Balfour Report was published in 1928. The Minister accepted the committee's recommendations with some reservations and commented:

> The Minister has no intention of suggesting the employment of one sex to the exclusion of the other on house management, but it does not appear to him that there are any functions involved which cannot adequately be performed by women of the right training and character.
>
> (Ministry of Health, 1938)

The report, after discussing the need for 'social education' for some tenants, had come to the conclusion that women were best fitted for this kind of social work and that the same person should have contact with slum tenants before rehousing and after. It rejected the Institute of Housing's view that there was no need to bring the rehousing officer into early contact with the slum tenant. While avoiding any rigid system it considered that the combination of 'social' work and rent collection or repair ordering was good as it gave a pretext for entry into the house, but it did not insist that all these functions be combined (CHAC, 1938: 28). The committee also directed attention to the need for these managers to have the right training as well as the right personality and listed 'sociological' subjects which they thought should form the basis for training, for example unemployment and other benefits, rent and wages, community centres. They went on to itemise the technical subjects which should also be included and concluded, 'Women trained on this basis would be of value to the housing department of a local authority and should be adequately paid' (CHAC, 1938: 29).

The decidedly cautious endorsement in the circular was clearly less than the Society hoped for though the emphasis on trained staff was helpful. Unfortunately by November 1938 the growing threat of war was already beginning to overshadow longer-term plans for the improvement of the service (*SWHM Quarterly Bulletin*, April

	1931 or earlier* 1934	1932	1933
Local Authorities	Bebington (1928) Bromborough Chester (1928) Chesterfield (1927) Cheltenham (1929) Kensington Leeds (1929) St Pancras Stockton Walsall West Bromwich Westminster (1930) Rotherham (1928) Hendon (1929) Norwich (1929) LCC (1931)	Brighton Cambridge Newcastle under Lyme Tunbridge Wells	Paddington Borough Council
Housing Associations etc.	Birmingham Copec Improved Tenements Association Church Commissioners Crown Commissioners Manchester Housing (1926)	Bethnal Green and East London Housing Assn Cambridge Housing Society Westminster Housing Trust United Women's Homes Ltd.	Aubrey Trust Liverpool Improved Homes Ltd. Shoreditch H.A.
			Kensington Housing Trust National Model Dwellings Co. Ltd. Newcastle on Tyne House Improvement Trust Ltd. Pilkington Bros, Cape Town

Bootle
Chelsea
Croydon
Lancaster
Lincoln
Tonbridge
Tynemouth
Bognor Regis
Brentford & Chiswick

1935	1936	1937	1938	1939
Local Authorities				
Westhoughton	Abingdon	Southall	Aylesbury	Battersea
	Hastings	Swansea	Fulham	Romford
	Mitcham	Winchester	Stirlingshire C.C.	St Albans
	Islington	Hemel Hempstead	Halesowen	Sowerby Bridge
				Uttoxeter
Housing Associations etc.				
	Chelsea Housing Improvement Society	Battersea H.A.	Messrs Clutton	Fulham Housing Improvement Society Ltd.
	Isle of Dogs Housing Society	Church Army Housing (Gateshead)	North Eastern Housing Association	
	Duchy of Cornwall			
	Leeds Housing Trust		East London	Port Elisabeth
	Oxford Cottage Improvement Society		Johannesburg	Sydney, NSW
	Paddington Houses			
	Swaythling Housing Society, Southampton			
	Thistle Property Trust, Stirling			
	Willesden Housing Society			

Figure 4.1 Local authorities and housing associations employing women managers, 1930s
Source: Compiled by author from SWHM minutes, reports and quarterly bulletins

1939b). Power considers that the Balfour Report on the whole favoured the Society's approach over that of the Institute. 'However, it failed to make clear cut recommendations on many key issues' (Power, 1987: 33).

THE EMPLOYMENT OF WOMEN HOUSING MANAGERS

In Figure 4.1 an overall summary is given of available information on local authorities and associations where women housing managers were employed in the 1930s.

The spread of employment was fairly slow compared with the total number of housing authorities; 1,716 in 1934 (*Municipal Journal*, 16.3.34). However, many authorities would not have had sufficient houses to justify the appointment of a housing manager; even by 1936 it was estimated that only 17 per cent of housing authorities had a housing department at all (Institute of Housing, 1936a).

In the 1930s women's scope in employment was generally restricted. In 1931, 38 per cent of all women aged 15–59 and 11 per cent of married women aged 15–59 were employed or seeking work ('Occupied' in census terms) (Hakim, 1979: 3). Women tended to be concentrated in typically female jobs, and Hakim (1979) has argued that over the period 1901 to 1971 occupational segregation did not decrease significantly. The housing field was not one which *a priori* was likely to be seen as typically feminine; for example, one of the related professions, surveying, had only recently become open to women. It was unusual for a woman to have authority over men, so women managers faced particular prejudice. Women were more acceptable in the social work role and, as we have seen, SWHEM had to fight to avoid this kind of pigeon-holing.

How should the progress, or lack of it, in women's employment in housing be assessed overall? It could be argued, from looking at the list in Figure 4.1, that the progress made was very creditable. Forty-eight local authorities were, by this list, employing women managers. But two factors have to be borne in mind. The first is that the list included some local authorities where women managers were employed under the 'dual system': where a woman managed the more difficult estates and a man the rest. The second factor is that there were 1,716 local authorities. Though some were small, some big authorities such as Birmingham, Manchester and the LCC were in the hands of predominantly male staff. (Manchester employed its first woman assistant in 1929 but in such departments women rarely reached the higher posts.)

So from the formative period of the 1930s women did not seem to be appointed to the largest authorities. On the other hand, in 1939 the *Quarterly Bulletin* claimed that 'Women housing managers serve metropolitan boroughs containing nearly ¼ of the population of London' ('H.G.L.A.', 1939). The London boroughs did have severe housing problems and were fairly prestigious appointments though many of the departments were still fairly small or had only limited powers.

At the point when they gave evidence to the Balfour Committee in 1936, the Society had 143 qualified members, 62 of whom worked for municipal employers and 72 for non-municipal employers. Power claims that 'by the late 1930s only 75 of its members were actually employed on council estates. Between them they were covering 35,000 properties, just under 500 properties each, and less than 5 per cent of the total council stock' (Power, 1987: 31). The Society's membership statistics, analysed in detail in the next chapter, show 375 members by 1939, but do not detail employment. The Institute of Housing had a membership of 261, 'not limited to officials engaged in local government nor to housing managers as such'. Women housing managers were facing considerable difficulties, as examples will illustrate.

DISPUTES OVER THE EMPLOYMENT OF WOMEN MANAGERS IN LOCAL AUTHORITIES

Leeds

In 1929 a woman housing manager and two assistants were appointed to look after older property in one area. The appointment of the women was attacked by Councillor O'Donnell (in the *Leeds Weekly Citizen*, on 7 June):

> Councillor O'Donnell strongly protested against this proposal, which he described as ridiculous and objectionable. It was a waste of public money to appoint these ladies to do work which could be done at less cost by clerks from the City Treasurer's Department. Rent collectors need not be university graduates. The fact was that these ladies were being appointed not only to collect rents but to do a little 'slumming,' to kiss the babies and quote Shakespeare and teach people how to sneeze by numbers!

The work in Leeds was started in an atmosphere of criticism and it seems that the circumstances gradually became more difficult. Leeds started to build up to a large slum clearance programme. In 1934 the Council appointed a Housing Director, Mr R. Livett, RIBA, who had held a senior post at Manchester (Ravetz, 1974). At a SWHEM council meeting Miss Marshall reported on her experience of these events:

> There is now one Housing Department at Leeds, under a recently appointed Housing Director who was formerly at Manchester. Miss Marshall and a male colleague work in the Rent Collecting section of the Department, which consists at present of twenty collectors and three trainees – men and women – and between them collect the rents of all the Council houses in Leeds. Until June 1934 Miss Marshall was responsible for a 'watertight' Octavia Hill Department, but now that is abolished.

From 1935 onwards mention of Leeds in the Society's records disappears.

The London County Council

Power (1987) gives a brief survey of the development of housing management in the LCC. They had appointed a Director of Housing in 1919 at a high salary. (The LCC had a Housing Manager for some years before that (LCC, 1908).) Building had largely concentrated on low-rise outer estates and these had been managed by resident superintendents. The model of management was similar to that used by Peabody and some of the other large trusts.

Power considers that the LCC were tied to the logic of large estates with one resident superintendent managing 2,000 dwellings. When changes of policy in the late 1920s increased the number of poorer tenants and renewed emphasis on flat-building in the inner areas, the lack of a coherent management system produced problems. The LCC therefore decided in 1930 to employ women housing managers with responsibility for door-to-door rent collection, repairs, cleaning, tenancy matters and court action.

> However, the LCC only sustained the intensive system for a few years and on a few estates, with the appointment of one woman manager and two female assistants in 1930. It reacted defensively and narrowly to the Government's report. [Balfour Report]
>
> (Power, 1987: 37)

The LCC argued that the Octavia Hill system was too staff intensive and expensive, but in fact the LCC had a very intensive staff ratio, and could certainly have integrated local management if it had wanted (Power, 1987: 38). Some of the other factors which caused the LCC to reject the 'Octavia Hill' approach and the employment of women are revealed in correspondence in *London Town*, the journal of the LCC staff association.

An article by 'Maria Brown', 'Slums and social service' (M. Brown, 1936a) advocated the employment of Octavia-Hill-trained managers, while stressing that estate management was and should be a career for both sexes. This provoked a violent response from an R. J. Fowler in the February 1936 issue. Under a title of 'Unpractical spinsters' he introduced some openly sexist arguments.

> In connection with repairs, a woman is usually at a disadvantage in executing minor repairs herself and dealing with work under her supervision . . .
>
> The basic argument used in favour of the Octavia Hill system is that women possess peculiar advantages for social service. Actually, the type of woman attracted to housing administration as a career is usually not the ideal conceived by 'Maria Brown', but a spinster with limited practical experience of household management . . .
>
> (Fowler, 1936)

The women answered back:

> With regard to supervision of repair work, most workmen carry out intelligent instructions willingly, and are much more influenced by the efficiency and application to duty of the officer in charge than by the sex.
>
> The fact that the second woman superintendent to be appointed by the Council left the service to take up a more congenial housing appointment which offered more scope for her ideals can scarcely be used as an argument against further employment of women as housing superintendents. There is no antagonism against men, but women would welcome the opportunity of working side by side with men under equal conditions in estate management as in other branches of the Council's service.
>
> (MacKenzie, 1936)

It is of interest that some of the letters subsequently written to *London Town* by women protesting against this decision were written under pseudonyms.

Liverpool

In 1935, Liverpool Housing Committee appointed Jean Thompsom (a SWHEM member and at the time managing property for Rotherham) to the post of Superintendent of Lettings. Later the appointment was rescinded by the whole Council on the grounds that she would be in charge of men (SWHEM minutes, 19.5.35). This sparked off a controversy which reveals many of the attitudes prevalent at the time.

Sir Gerald Hurst in Parliament asked the Minister of Health 'whether he is aware of this incident'. The Minister (Sir H. Young) replied:

> my attention has not previously been drawn to this matter. The selection of any officer in any particular instance is within the discretion of the appointing authority, but in the last two annual reports of the Ministry of Health the attention of the local authorities was drawn to the importance of the appointment of properly qualified officers for house management, and I propose to include some further observations on this issue in an early circular letter.
>
> (*Municipal Journal*, 17.5.35)

(Sir Gerald Hurst was the father of Margaret Hurst, later Margaret Hill, who was trained with SWHEM and later became the first woman housing manager in South Africa and the first housing management adviser at the Ministry of Health.)

Eleanor Rathbone, MP, who had been a member of Liverpool City Council for 25 years, wrote

> Many councils scarcely recruit women at all except for routine clerical work and for special posts in the educational and health service which only women can fill. Even when there is no formal bar against women and girls as competitors they are often in practice excluded.
>
> There is no justification for a sex impediment in filling municipal posts. . . . Many administrative posts call in fact for qualities in which women are, by tradition, supposed to excel, such as an intuitional knowledge of character, tact in handling subordinates, thrift and economy, meticulous attention to detail, and, above all, knowledge of and imaginative sympathy with those who are suffering from diseases, poverty, bad housing conditions, unemployment etc. It is not a mere metaphor to say that local administration is like housekeeping on a large scale and calls for much the same qualities.
>
> (Rathbone, 1935)

Despite the pressure, no change was made at Liverpool.

Ideally women's work?

In September 1935 an article by Emily Murray, a senior member of the Society, entitled 'Housing management is "ideally women's work"' appeared in the *Municipal Journal*. It argued:

> Trained management is proving itself, and those who realise its value generally recognise that it is ideally women's work. The working-class woman conducts the business of the home in her husband's daily absence. The woman manager starts with a working knowledge of the difficulties inherent in housework (a bond of sympathy at the outset) and, in pursuit of her duties, she will find many points of contact and many opportunities of enlarging the scope of her work to the mutual advantage of tenant and landlord.
>
> (*Municipal Journal*, 6.9.35)

This article provoked a response from Ernest France, 'Chairman of the Rent Collectors and Investigators, Manchester':

> It would be a misfortune if, in our anxiety to obtain the best results, we overlooked the fact that the bulk of the pioneer work has been done by men trained in the hard school of experience. Many of them, to the writer's knowledge, are ready to endorse the methods and principles of the late Miss Octavia Hill.
>
> Those of us who have had this experience have found that men have certain advantages in estate supervision particularly in regard to disputes between tenants; and the argument that only a woman can know what is right or wrong about a house is one which would be disputed even by many working women. It is probably true also that a housewife would more readily accept a suggestion or reprimand from a man.
>
> (France, 1935)

Margaret Miller (then Secretary of the Society) replied

> Without wishing to belittle in any way the work done by men in the field of housing management during the post-war period, there can be no doubt that it was a woman, Miss Octavia Hill, who discovered the right approach to house property management as long ago as 1864, and it is by women that the principles of expert and enlightened management have been developed since that date. This is not to deny that men have successfully adapted and applied

Miss Hill's methods and that men may make a valuable contribution to the solution of housing management problems in the future.

But we do claim that the profession of housing estate management is and will remain one for which women, by reason of character and temperament, are specially well adapted and in which, given the requisite training, they are able to do specially valuable work.

(Miller, 1935)

The correspondence continued for some weeks and exposes the Society's dilemma. It was all very well to say that housing, like other occupations, should be open to both sexes, but the fact of the matter was that women were at a grave disadvantage when it came to employment in local authorities. In order to overcome this disadvantage and to give themselves a special edge, they used arguments about a 'special aptitude' of women for the work which implicitly reinforced the common idea that women had inherently different mental capabilities from men. This argument, based on sex role stereotyping, could easily be turned against them.

The experience of interviewees

In the case of new appointments of Octavia Hill staff, most interviewees stressed that what was important was that the Council had made the decision and were firm about it and then the officers accepted it.

The officers thought we were rather odd but they were quite nice to us. Of course it had been a decision of the Council to have us and they knew that. They didn't really understand what we were trying to do.

(Interview, Society member)

Thus at Lancaster 'It was accepted that the housing department was run by women, full stop' (interview, housing manager who trained at Lancaster). A number of 'Society' offices were all-women offices as regards the housing management staff and remained so for many years. In other cases, particularly where a housing department was formed after rents had been collected by another department, these male rent-collecting staff would be taken into the staff of the new 'Octavia Hill department'. One or two of the interviewees had worked quite happily in such offices where they might be supervising or training male staff and stressed the wide practical experience that these staff often had. But another felt that 'If a woman was doing a job she was accepted . . . housing management itself was new. . . . Octavia Hill had done it. . . . But the idea of a man training under a woman was not on.'

So the experience of interviewees in encountering discrimination was influenced by the arrangements of the office they worked in. But anyone looking for a new post or opening up a new office had to prove over again that women were capable of doing this kind of work. Though there were some examples of women managers who supervised male staff, in many cases this was avoided by having 'all female' departments.

EMPLOYMENT IN HOUSING ASSOCIATIONS

Figure 4.1 shows that there was also a steady increase in the number of appointments to new housing associations during this period. The associations employing women seem to have been the smaller ones, often formed by people who were influenced by 'Octavia Hill' ideas. Some examples are given below.

The Improved Tenements Association had begun from work started in 1899 by Octavia Hill. Miss Dicken had managed the property for her and continued to manage after her death. The Association dealt in reconditioning work only and continued to employ women housing managers after its amalgamation with the Rowe Housing Trust ('E.M.', 1934).

Birmingham Copec House Improvement Society started from a conference on Christian Politics, Economics and Citizenship held in Birmingham in 1924. Great concern was expressed about the conditions in back to back houses and a group of people from the churches continued this concern by forming the Society in 1925 (Fenter, 1960: 1–5). As the number of houses owned by the Society increased

> it was recognised that a full-time paid worker would be necessary . . . and early in 1926 the Society interviewed Miss F. M. Fenter, a university graduate who, after gaining the Social Study Diploma at the Settlement, agreed to go to London for a period of training with Miss Jeffery and with Mrs Barclay and Miss Perry of the St Pancras House Improvement Society.
>
> (Fenter, 1960: 13)

Liverpool Improved Houses Limited was formed in 1937 'to help the poorest tenants, left in the overcrowded districts, where their needs have hardly yet been touched by public effort and the condition of the houses has become worse rather than better during the last ten years'. By 1928, they 'have arranged for all their property to be managed on the Octavia Hill system, and have also undertaken the management of similar property belonging to the Marquess of Salisbury' (Liverpool Improved Houses Ltd, 1928). Liverpool Improved Houses became important as a training office for the Society and made a strong impact on a number of trainees (interviews with society members).

A specific example of women's concern about housing was the formation of two associations concerned with single women. *Women's Pioneer Housing Limited* was formed in 1921 as a co-operative society with the aim 'of providing attractive flats for women who must make their own homes' (Women's Pioneer Housing Ltd, 1935). After some financial difficulties in the 1920s it was able to expand more in the 1930s though still on a modest scale; 522 flats by 1935 (ibid.). *The Over Thirty Association* (now Housing for Women) was originally founded in the early 1930s to assist older women in obtaining employment. The immediate response was to start a hostel but, as time went on, it became clear that middle-aged women wanted self-contained housing so the association was gradually drawn into work in converting houses for flats, though most of this was after 1945 (Over Forty Association, 1981, 1983).

From interviews with women managers and from correspondence in the societies' journals it would seem that most of the housing associations employing women managers were like the ones quoted above: fairly small, often founded by groups of people who had some common voluntary activity for some time and continued to be so. Many of these associations were linked with a tradition of voluntary effort in housing, of which SWHEM was itself a part. Where women had been employed from the beginning it seems to have been easier to continue women's jobs. But the largest housing associations, for example Guinness, Sutton and Peabody, had predominantly male staff and were more likely to support the Institute of Housing.

CONCLUSIONS

In the period 1932–39 public and social housing work expanded. In general employment women tended to be concentrated in typically female jobs and housing was not one which *a priori* was likely to be seen as typically feminine. Women were accepted in the social work role but the women housing managers were keen not to be limited to this. Eleanor Rathbone's article indicates that local government was not in itself particularly receptive to women's participation except in routine clerical work. The high level of unemployment during the period was sometimes used as a reason for not 'taking men's jobs'. There was often reluctance to put women managers in charge of men. So women managers had an uphill battle on their hands.

As housing organisations began to take on the task of slum clearance, more intensive management work was seen as necessary and women were considered to be suited to this approach. The association of women with a particular type of 'Octavia Hill' management could be turned against them if that type of management was seen as oppressive and interfering – or just too expensive. The controversies in the *Municipal Journal* and *London Town* demonstrate both the degree of prejudice which the women were facing and the dangers in arguments about women's 'special aptitudes' for this type of work.

5 Men, women and professionalism in the 1930s

In face of the kind of opposition described in the previous chapter, women needed support and help in finding employment opportunities. How did the Society organise itself to fulfil this role and how effective was it? Why did this period see the development of two professional organisations, one for men and one for women? The focus of this chapter is on the Society and the way in which it built up the relationship with members, as well as campaigning on their behalf.

THE BEGINNINGS OF THE SOCIETY

The first Annual General Meeting of the Society of Women Housing Estate Managers was held on 29th October 1932. From this time forward the Society usually had a London office, a paid secretary and, after 1934, an assistant secretary and sometimes a paid clerk as well.

In 1935–36 the Society decided to apply to the Board of Trade for a certificate of incorporation. Parker Morris had been an important influence in pointing out the legal difficulties of not being a corporate body and had given much help and advice over the process of incorporation (SWHEM, 1937). The Society of Women Housing Managers was incorporated on 8th September 1937 as an association limited by guarantee. Among the objects of the Society as registered were:

(A) To follow and extend the principles initiated by Octavia Hill for the management of house property.

(B) To provide a central organisation for persons engaged in or connected with the profession of Women Housing Managers trained in and following the principles of Octavia Hill, and to raise the status and to promote and encourage the interests of the said profession.

(C) To improve the technical and professional knowledge of members of the said profession by the provision of a library, the arrangement of meetings, lectures and discussions, the organisation of study tours.

(SWHM, 1937: 5, 6)

There were two classes of membership: Fellows and Ordinary Members. The first meeting of the incorporated Society was held at the London School of Economics in November 1937, with 101 Ordinary Members, Honorary Members and Associates present (SWHM, 1938: 4).

The junior organisation

This organisation came formally into being on 3rd December 1938. Its purpose was to enable younger members and students of the Society to meet and discuss matters of general and professional interest. It consisted of four regional groups and a central executive. The 1938–39 Annual Report stated that each group had had several meetings and speakers, including senior members of the Society and other experts on technical subjects (SWHM, 1939: 6).

RECRUITMENT AND TRAINING

Like the AWHPM and the Octavia Hill Club before it, the Society spent a considerable amount of time and energy on recruiting suitable students and training them. A training secretary was appointed (SWHEM, 1933) and considerable discussion of training is recorded.

The Society tried to promote awareness of housing management as a career by contacting schools and universities, by getting material on housing management included in careers guides, and through their policy of public speaking to as wide a variety of groups as possible. Applicants had quite often had some personal contact with a member of the Society through these activities or had spoken to somebody who knew about the work. For example, headmistresses were mentioned in a couple of cases and one of the tutors at Somerville was instrumental in introducing two or three students to the idea of housing management (interviewees).

All students had to be interviewed by members of the training committee before being accepted for training (*SWHM Quarterly*

Bulletin, April 1939a: 1). The interview, which often seems to have been before quite a large committee, was sometimes a daunting process and tended to make quite an impression on would-be trainees. A number would echo the view that it was 'quite ghastly' (interview, member of the Society) but some appreciated the care given.

> 'Well I can remember being very interested in the sort of quality of the interview which I had; it was very meticulous, it was very very searching. The thing that I was impatient with . . . was their very rigid rules about training, in the sense that they interviewed me and said yes we think we will accept you for training but we can't place you until goodness knows when . . . you see this was the first problem that I think the Society was always up against – that they were just too tiny and too dedicated, if you like, to expand.'
>
> (Interview, Society member)

A degree was regarded as desirable but not essential. Because a higher standard of education was required there was a certain class bias.

> 'It was a bit [middle class] wasn't it? I think, you see, pre-war mostly the people who had the educational background required were people of upper middle class on the whole, or middle class, they weren't all "ladies" . . . but in asking for the educational background in those days you were picking a social type also. . . . In local government offices you did get paid and students' pay got better; but you were expected to "rough it".'
>
> (ibid.)

Students had to pay an entrance and training fee to the Society of 20 guineas. In addition they had to meet lecture and examination fees which would vary according to the examination taken (SWHEM, 1934a: 11). They were often not paid, or paid very little, except in municipal posts where the rates were not high but training fees might be paid (SWHEM, 1934a: 11). The financial burden of training implied that most students or their families had to have some private resources. However some students came from rather less well off homes and made considerable sacrifices in order to train, for example saving money from a previous job to pay for the training period (interviewee). Some effort was made by the Society on behalf of those who were less well off. In 1939 the Octavia Hill Scholarship Fund was set up to help them.

Theoretical training

The provision of adequate theoretical training and time for study remained a problem. The syllabus was still provided by the Chartered Surveyors' Institution and tuition by the College of Estate Management, but there was some dissatisfaction with both of these. The 1930s syllabus contained mainly technical subjects, together with economics, law and government, but its spread could cause difficulty to students:

> The mixture of legal and scientific subjects is attractive to some students and difficult for others. The same person cannot as a rule shine at valuations, mathematical subjects, drawing plans, grasping legal points, and writing good reports, and some candidates will find it necessary to take longer over the course than others.
>
> (SWHEM, 1934a: 12)

In July 1933 the Training Committee discussed the desirability of having two classes of certificate, one for managers and one for assistants (SWHEM minutes, 8.7.33: 14). This suggestion was not accepted but the possibility of having a less exacting qualification was discussed from time to time and carried out after the war. Discussion of the syllabus continued (*SWHEM Quarterly Bulletin*, January 1936a). There was particular dissatisfaction with the lack of social services content. The tuition provided by the College of Estate Management was advertised as being whole-time, evening class or postal (*SWHEM Quarterly Bulletin*, April 1936) but this meant that students outside London had to study by correspondence, with all the attendant disadvantages. Comments on training in the Balfour Report, advocating better knowledge of the social services, led to a thorough revision of the syllabus to include subjects such as Family Income and Social Services (SWHM, 1939: 7).

Practical training

Practical training continued to be given considerable importance. There was some discussion as to whether training in two offices should be compulsory and this seems to have been favoured by Council. But it was held that making it compulsory rather than desirable for municipal trainees to work in a non-municipal office 'would fatally hamper the best prospects for the future'. The Training Committee established a list of offices recognised for training students (SWHEM minutes, 6.5.33). The Training Committee regularly reviewed the progress of students in practical training and if their experience or standard of work was not considered satisfactory they would be referred for a longer period of training and not recommended for full membership (SWHEM minutes, 16.11.35 onwards). It also reviewed the practice of the training offices (*SWHM Quarterly Bulletin*, April 1939a):

> 'They kept very close contact. We had six-monthly reports. It was very intensive training both from the people who were doing the training and the trainees, and the training manager had to send in regular reports and you got interviewed every now and then.'
>
> (Interviewee who started training in 1935)

Nevertheless students were sometimes critical of their training offices. But the practice of training in two offices meant that if one office was not congenial the other one might be. In the process of training, interviewees often came into contact with people whom they later regarded as having been very influential in their career.

The functions of the training system

The role of training in providing coherence to an occupational group has been widely recognised, particularly in the literature on professions and semi-professions (a term used by Etzioni, 1969 to define jobs like teaching and social work which lack some of the characteristics of true professions). Millerson (1964), for example, includes training and education among six essential features of a profession. The present study is not concerned with measuring exactly how far along the continuum of professionalisation the Society had proceeded at each stage. What is valuable from the literature on professions and semi-professions is the understanding of the role that training and examinations play in occupational development.

Wilensky (1964) constructed a table which summarises the development of 18 occupations and identified eight steps in the process of professionalisation, the establishment of training schools being one step. The literature indicates that training would both increase the status of the Society and make the group more cohesive. Sandle (1980) provides one of the few discussions of professionalisation in the housing context, though this was related to the Institute of Housing in the 1980s. She demonstrated that, despite claims by the 1980s Institute of Housing that a 'housing professional' existed, this concept was in fact hard to identify. She came to the conclusion that housing was a long way from being fully professionalised, because:

1 The housing professional has not been clearly identified.

2 It has not fully identified its own body of knowledge and skills.

3 It has totally ignored the socialisation of its professionals and has not developed behavioural norms that foster a professional identity, so that its members are subjectively conscious of themselves as being part of a community.

(Sandle, 1980: 107)

The evidence presented here about training procedures and the interviewees' comments suggests that the Society did clearly identify the housing profession, socialise its members and develop behavioural norms. Its members were subjectively conscious of themselves as being part of a community.

The training scheme had its disadvantages. The main one was that selectivity and close personal supervision meant that the supply of training places was very limited and students could only be trained slowly. Interviewees felt that this was a major limitation on the growth of the Society in the 1930s and in the post-war period. The number of students during this period is given in Table 5.1. Expansion was painfully slow. Numbers were limited by the lack of finance for housing training. Though the Balfour Committee considered training, it limited itself to making recommendations about the content of training and the need for adequate payment of trained women staff (CHAC, 1939: 29).

THE SOCIETY'S VIEW OF HOUSING MANAGEMENT

At the heart of the training was the Society's distinctive view of housing management and the role of the woman manager. The Society's evidence to the Balfour Committee, outlined in the previous chapter, was based on this. The official Society view is more fully described in *Housing estate management by women*, published in 1934 and subsequently reissued many times. This discusses the role of the landlord and says that Octavia Hill realised that 'the supply of living accommodation to the poor could not be regarded primarily as a profit-making business without disastrous effects on health and civic life'. But 'Octavia Hill did not devise a system suitable to the widespread public control and ownership of today'. Her approach had been to demonstrate the immediate advantages to

Table 5.1 Membership of the Society* 1932–39

	Oct. 1932	Aug. 1933	1934	1935	1936	Aug. 1937	1938
Full members	86	102		122	148	164	159
Fellows (after incorporation)	–	–		–	–	–	26
Professional student members	28	29		47	58	54	60
Honorary members	–	–		–	24	27	30
Associate members (created at AGM 1932)	–	33		43	66	73	63
Totals	114	164		212	296	318	338
Students qualifying:							
PASI (Finals)	2 (referred)	1		1			
Inter.	1						
WHPM Certificate	2	12		26	14	19	13
Sanitary Science Certificate	3	2			2		
BSc in Est. Management University of London							
Final		1			1	2	1
Inter.				2	1	1	
University Diplomas in Public Administration					2		

Source: Society of Women Housing Estate Managers*
Annual Reports 1932–1939

* Society of Women Housing Estate Managers to 1937
Society of Women Housing Managers from 1937

be gained from improving the management of working-class houses. But

> There is now a general acceptance of the principle that if the provision of living accommodation for those who are too poor to bargain freely and obtain value for money is to be undertaken for private profit, the interests of the tenant must be safeguarded by the State.

(SWHEM, 1934a: 3,4)

The Society argued from this that the 'business' and 'social welfare' aspects of housing management are not independent spheres of activity.

The second principle outlined is that of 'homes not houses': that the need for a home for a family should govern both the design of new property and the improvement of older property, rather than the profit motive. Tenants may sometimes need to be 'roused' to their responsibilities. 'From this evolved the idea of sympathetic management, dealing with individuals instead of with tenements in the mass' (SWHEM, 1934a: 14). The need for trained staff to deal with tenants in this way is emphasised.

The scope of the work is outlined. This includes rent collection and rent fixing, dealing with arrears, and it is stressed that this must be appropriate to the circumstances of the family. Advice to tenants on money matters and help with employment or contacts with officials is also mentioned. Maintenance of property is discussed, including both repairs and disinfestation. It is pointed out that in private work this may involve direct employment of building operatives, whereas in local authorities it would usually be carried out by another department. The keeping of accurate accounts and balancing them is stressed. Selection and placing of tenants, court work and records, reports and committee work are also part of the job. Finally, the broader range of a manager's work, in new development and in care for open spaces and community facilities, is discussed.

Jean Thompson's 1931 paper 'The administration of municipal housing estates' can be taken as representing the more progressive views within the Society. Thompson argues that there is an urgent need 'for thinking out and clearly formulating a body of principles to govern municipal housing administration' because the previous experience of the private landlord provided no such principles to go on. Such principles do need to take account of the human and social as well as financial and technical aspects of management. Thompson discusses the tension between the 'social aspect' of housing and the 'municipal trading aspect', and says that it is disastrous to ignore either.

As a basis for the new approach, local authorities should recognise that '*the majority of tenants will respond to efforts made to improve their environment* but that the extent of the response depends very considerably on whether the estates are well-managed or not'. In actual operation, the principle will demand the setting up of a properly constituted housing department with a responsible manager, because splitting up the functions between several departments results in a lower standard of efficiency with lack of responsible management.

Thompson discusses rent collection and arrears, taking into account the fact that tenants often feel that the Council can afford to lose money more than the private landlord; but, if arrears are simply allowed to mount, the development of collective ownership in housing will ultimately be retarded. On the other hand, she points out the difficulty of taking action on arrears in areas of widespread unemployment. The stress on careful selection and placing of tenants and on maintenance work is repeated. Thompson's view is that tenants should be encouraged to report defects in their housing because the local authorities' maintenance work is part of looking after its own asset. The importance of advisory work on new building needs, and of a co-ordinated approach to housing, is stressed. Thompson particularly emphasises the need for selection of tenants to be done by officers rather than councillors to avoid abuse, dealing with a problem that was to surface many times in housing.

Thompson's paper represents an attempt to begin to think through principles of the management of public housing at a level which is missing from most other publications of the time. How far did the practice of housing management by Society members differ from that in other organisations?

The Balfour Committee showed an enormous variety of housing management practice, with some local authorities reaching a rather poor level. Its papers and visits also reveal a strong influence still pervading from the Public Health Movement, especially because of slum clearance. Like the Society, the Institute of Housing and the AMC emphasised the need for better housing management. The Society focused on the delivery of the service to the tenant and the need to integrate the 'business' and 'social' sides of the work. In general the Institute of Housing and AMC papers concentrate on 'business'. 'Welfare' appears rather as an 'add-on' extra, directed especially towards ex-slum-clearance tenants. They reveal a large-scale bureaucratic approach. Society interviewees felt strongly that the actual *practice* of management differed considerably (Institute of Housing, 1936a; Association of Municipal Corporations, 1937).

Studies of housing management in the 1930s as yet available do not contain the detail and range of evidence which would enable a firm judgement to be made. Given the lack of consultation with tenants in the period, it may be impossible ever to get such evidence. The documents and records of the Society plus the evidence from interviewees indicate that considerable efforts were made to ensure effective management which could benefit both owner and tenant.

The Society's definition of the identity of the woman housing manager was the other distinctive contribution. As we have seen in Chapter 4, other organisations, if they accepted women at all, were prone to relegate them to the role of welfare worker. The Society strenuously resisted such efforts, spelling out clearly its idea of a full management role for women. This was of key importance in encouraging women into the career.

Meetings and conferences

Figure 4.1 showed that the places where women housing managers were appointed were scattered all round the country. Many were in fairly isolated appointments or in offices with a small number of staff. The Society provided an opportunity to meet other women managers. The system of training meant that, by the time they had their first appointment, qualified women usually knew staff in two or three offices; so each person built up a range of contacts. For many members an important opportunity to renew those contacts was provided by the meetings of the Society, in particular the Annual General Meeting and the Annual Provincial Conference.

'We used to go to Annual General Meetings, that was the great thing and everybody went – they were on a Saturday so that everybody could go. . . . It was one of the nice things about the Society . . . that you all knew each other and probably had contact with each other at one time or another.'

(Interview, Society member)

The Provincial Conference and the AGM in London were important and some areas would have a local conference too.

The style of the Society, according to interviewees, was informal and meetings were held without undue pomp and circumstance. This meant that members built up social networks. The importance of such networks for women has been discussed by a number of writers. Ryan comments on women's voluntary associations in one neighbourhood:

> Most were congregations of peers; members of similar age groups, occupations and ethnic backgrounds. Most rejected a rigid governing hierarchy and condescending manners. . . . The association relied on informal but expansive social ties, a voluntary network of like-minded individuals, as its organisational machinery and political leverage.
>
> (Ryan, 1979: 68, 69)

SHM interviewees often used the phrase 'like-minded' to describe the membership of the Society. The significance of such organisation for the Society was not only that it could operate effectively as a campaigning body; it also provided considerable support for its members.

PUBLICATIONS

The Society had the task of making its existence and work known to as wide an audience as possible, in order to gain general support and opportunities for employment. The Society followed the pattern of AWHPM and produced a range of publications.

The *Quarterly Bulletin*

Publication of the *Quarterly Bulletin* began in 1933. The first issue stated:

> It is hoped that the *Quarterly Bulletin* will be a means of keeping all Members in touch with the Society and its activities; but the aim of the *Quarterly Bulletin* is not to be merely a list of happenings; it is hoped that in the future it may become a useful organ for the exchange of ideas and experiences. With these objects in view, we do urge members to send contributions – items of news, articles bearing on some particular phase of the work, or of general interest.
>
> (*SWHEM Quarterly Bulletin*, April 1933)

In the middle years of this period, the content settled to that of an editorial (usually short), articles and reports (at times in some detail) of conferences and meetings attended by Society members. Members' News included short reports from members on their work, notes on publications, notices of forthcoming meetings addressed by members, and a note of members' appointments.

The bulletin seems mainly to have been a means of informing and supporting members, though Parker Morris had argued that it should be used for publicity (*SWHEM Quarterly Bulletin*, January 1936b).

Housing estate management by women

In 1933 the Executive decided to publish 'A short summary of the Society's aims, together with a list of the personnel of all Committees with the members' university and professional qualifications. . . . Various members were deputed to work on this booklet' (SWHEM minutes, 22.1.33). In time, production was completed and the pamphlet was circulated to all the major national, and some regional, newspapers. All the women MPs received copies. *Housing estate management by women* (SWHEM, 1934a) was to remain the backbone of the Society's publicity for some time, and was reprinted as needed.

Other publications

The Society published its Training Scheme partly, it seems from the content, for propaganda purposes. It followed the precedent set by AWHPM and the Octavia Hill Club of getting articles published in other journals and sometimes reprinting them for its own use. For example, *A day in my official life: housing estate manager*, by Jean Thompson was published by the Institute of Public Administration (Thompson, 1935a). An article in *The Times* 'Careers for girls: women property managers' (*Times*, 1939) was used as a reprinted leaflet. The output of such leaflets was not, however, quite as intensive as around 1930. It seems that some of the earlier leaflets were still in use. Parker Morris had urged that Society members continue to contribute to other journals, including women's magazines, and this did continue, with articles in the *New Statesman*, *Women's Magazine* and *Public Administration* (SWHM minutes, 18.11.39).

CONTACTS WITH OTHER ORGANISATIONS

It was important to the Society to keep in touch with general women's organisations which might take an interest in housing. For example, the National Council of Women was 'the co-ordinating body for a number of women's organisations' (Barrow, 1980) and the Society was represented on it from 1933. In 1933 the Executive recommended 'that application be made to the National Council of Women for the setting up of a sectional committee on housing' (SWHEM minutes, 8.7.33). Resolutions were also submitted. In 1936 Miss Upcott and Miss Galton submitted a resolution to the National Council of Women

> To urge on Housing Authorities, Public Utility Societies and private owners the necessity of providing, in connection with re-housing, an adequate minimum of square yards per family of play space for young children adjoining the buildings and of play space for older children near to those buildings.
>
> (SWHEM minutes, April 1936)

In 1936 another resolution was framed for inclusion in the National Council of Women agenda:

> That the National Council of Women, having previously urged local authorities to adopt the Octavia Hill system of management on their housing schemes, especially in reference to slum clearance, wishes now to call attention to the regrettable tendency of certain local authorities to make use of women solely in low-grade, specialised welfare posts, as distinct from employing trained women in full administrative control.
>
> (SWHEM minutes, 14.5.36)

SWHEM helped to set up the women's advisory housing committee described in Chapter 4. In general, SWHEM seems to have been willing to co-operate with other women's organisations on demands connected with housing or employment. The Society was affiliated to the Women's Employment Federation (SWHEM minutes, 19.5.35). An approach in 1935 from the organising secretary of the Over Thirty Association to raise the question of housing accommodation for single working women was considered by the Council. It was decided to inform her that SWHEM 'viewed with sympathy any endeavour to meet this need' (SWHEM minutes, 16.11.35).

It was also important to maximise opportunities to meet employers' representatives. For example, the National Housing and Town Planning Council (NHTPC) was a non-political body concerned with the improvement of the living environment, with a membership from local authorities, housing associations, building societies, etc. (*Housing and Planning Review*, 1981: 3). SWHEM was represented on it from 1933.

The relationship with NHTPC seems to have been quite close, with reports of its conferences appearing at regular intervals in the Society's bulletin (*SWHEM Quarterly Bulletin*, January 1936a, January 1937). Again the link was used to raise the issue of women housing managers and make new contacts at meetings and conferences.

There were similar links with a number of other organisations concerned with housing, such as the Royal Sanitary Institute, the Institute of Public Health, and the Garden Cities and Town Planning Association. Representation to the Charity Organisation Society was also agreed at the first meeting. The Society sent representatives to the Mansion House Council on Health and Housing and several members attended their meetings (SWHEM, 1933).

The representation of SWHEM at official government committees has been discussed in Chapter 4. The Housing Centre was another body with which the Society started to co-operate over this period; it began to publish a notice of their meetings in 1938 – the two organisations were housed in the same building so co-operation was quite convenient (*SWHM Quarterly Bulletin*, January 1938).

The Institute of Housing Administration

In its contact with other organisations, SWHEM usually gave and received support and avenues for further publicity. Its relationship with the Institute of Housing was more problematic. 'In 1931 the idea of forming the Institute originated amongst a group of municipal housing managers in the Midlands who had been in the habit of holding informal meetings to discuss problems connected with management' (Hort, 1934). The inaugural meetings were attended by several members of SWHEM and Miss Moor, Miss Fenter and Miss Hort were members of the original executive committee of twelve. Wallace Smith, estates manager of Birmingham, was very much the moving spirit of the Institute and was elected president each year until 1946 at least.

Initially membership of the IHA and SWHEM was not felt to be incompatible. Miss Hort wrote an article for the bulletin in which she argued for 'the fullest possible co-operation and harmony between the two bodies' though she realised that this would require 'tact and forbearance' (Hort, 1934). By April 1934, the *Quarterly Bulletin* reported signs of strain:

At the informal meeting of the members on Sunday morning the attitude of the Society to the Institute of Housing Administration was discussed. A paper written by Mr. Wallace Smith, of Birmingham, was discussed, in which he stated that in his view women should only be employed on housing estates as social workers, to 'work up' the bad tenants; the number of these he estimated to be 10%. Miss Alford said that we might accept this theory as reasonable and practical, but only when and if an estate contained a low proportion of bad tenants; even then, it would make difficulties both for the tenant and the social worker, as all the neighbours would know that when the latter called on a tenant, it was because the tenant was considered below standard. On an estate run on Octavia Hill principles this difficulty need not occur. If the estate contained a large proportion of tenants needing special supervision, this theory of management would be quite impracticable.

It was felt by those present that the Society should wait for future developments, and that no definite action need be taken at present towards the Institute. A resolution was passed to this effect, individual members being still free to join the Institute if they wish to do so.

(Gold, 1934)

Opposition of some IHA members to the Society's views hardened:

Miss Philipp reports from West Bromwich that she attended a meeting of the Midland Branch committee of the Institute of Housing Administration on Tuesday, June 18th. The meeting was called to discuss a resolution from the London committee of the Institute to the following effects:

(a) That the frequent reference in the Ministry of Health circulars and reports to Women Estate Managers is prejudicial to the interests of male members.
(b) That the minimum remuneration of Women Property Managers referred to in the Ministry of Labour circular 'Choice of Careers Series No. 4.a.' dated February 5th, 1932, is not being paid in many instances.

Miss Philipp spoke on the resolution, saying, among other things, that she would suggest that the Institute ought to be pleased that the Ministry stressed the importance of appointing 'fully qualified housing managers', as she was under the impression that the Institute was formed to improve the standard and status of housing officials throughout the country – to this the rest of the committee agreed. She suggested that the Ministry probably stressed *Women*

Housing Estate Managers simply because our Society was the only one having a training scheme; and that when the Institute has a training scheme the Ministry should be informed of the fact. She told them that the Society had passed a resolution that fully qualified managers should not apply for posts advertised at low salaries.

After an interesting discussion the committee passed a resolution to the effect that the Midland Branch of the I.H.A. did not consider the reference (as above) detrimental to the male applicant, providing that the woman was fully qualified (including examinations); and urging the executive of their Institute to hurry on the preparation of the training scheme and to bring the existence of the I.H.A. to the notice of the Ministry. One much applauded remark, noted by Miss Philipp, was to the effect that it was impossible for any person, male or female, to be *fully* qualified even in three years.

(*SWHEM Quarterly Bulletin*, October 1935b)

Wallace Smith, who was president of the Institute and had great influence as the manager of a large authority, was a focus of the opposition to women managers. An article in Municipal Journal described the office organisation and administrative work of the Birmingham Estates Department under the management of Mr Wallace Smith.

For rent collection the author does not oppose the Octavia Hill system, but personally prefers men as collectors and women as social workers. . . . Experience leads him to regard women as most capable of visiting prospective tenants but less satisfactory in visiting the installed tenants, yet the latter are ones for whom a woman's intelligent co-operation is needed.

(*Municipal Journal*, 20.4.34)

Such statements were hardly likely to have endeared him to the members of SWHEM. We read in the minutes of January 1936

That in view of the facts which have been brought to light Council is of the opinion that membership of the Institute of Housing Administration is rapidly becoming incompatible with membership of the SWHEM. Members of the latter body are, therefore, asked to consult together as to the most effective moment to resign in a body from the IHA.

(SWHEM minutes, 3.1.36)

The members of the Society who were also members of the Institute met and decided they would resign (SWHEM minutes, 3.5.36). From that time on the Institute was a very male-dominated organisation. For example, the 1936 Council, the first recorded, was all male and the first woman executive member was not elected until 1939. One woman member of the Institute Council remained the norm until 1948 (Institute of Housing, 1956). There does not seem to be evidence of any further co-operation between the Society and IHA in the period up to the Second World War.

PUBLIC SPEAKING

The Society encouraged its members to go out to give talks to organisations of all kinds and to schools and other educational institutions. This work was so diverse that it is difficult to summarise it.

Even the relatively new medium of broadcasting was used. 'Women's Management has also been the subject of B.B.C. talks and has been introduced into films, conspicuously that produced by the Under Forty Club' (SWHEM, 1934a: 3). Several interviewees had heard about housing work in this way: 'I heard Miss Y speak at an old girls' do;' 'I heard a talk at school by a member of the Society.' This aspect of the work was regarded seriously by the members of the Society and there was a suggestion that training could be provided.

CONTACTS WITH EMPLOYERS

Initiating contacts with individual employers, responding to requests for information from them in a very positive way, and arguing the case for the employment of women were obviously key activities for the Society. In 1934 it was resolved that where enquiries were made by potential employers, if they were in London the Secretary should try to obtain an interview, if in the provinces a member and the enquirer should be encouraged to see round some of the estates managed by members (SWHEM minutes, 10.9.33). Considerable time was spent, both by the committees of the Institute and by individual members, on these contacts since showing enquiring employers estates already managed by women was obviously regarded as a good selling point. Besides encouraging the employment of women managers, the Society publicised vacancies to its members and acted in an almost trade union role in negotiating salaries. Such negotiations were important in maintaining status for the women managers and avoiding their relegation to the welfare ghetto. The Societies' negotiations with Chelsea give an indication of the level of detailed involvement:

> Chelsea. Except for one or two minor alterations, the Chelsea Council accepted the conditions of service drawn up by the Sub-Committee of S.W.H.E.M. and the post of Manager was circulated on the agreed terms to members of S.W.H.E.M. only.
>
> (SWHEM minutes, 14.7.34)

Miss Jean Darling was appointed as Manager (*SWHEM Quarterly Bulletin*, April 1934b) and was influential in the Society. Even in 1948, Chelsea was still 'almost an all-woman office' (interviewee).

Channels of influence

In general, people are more easily persuaded to try something new if they are introduced to it by someone of status whom they respect (see, for example, Argyle, 1967: 30). Since men usually had the power at the time, the help of men who had become convinced of the value of SWHEM's work was crucial. It is likely that recorded instances are only the 'tip of the iceberg' of personal influence because such contacts are likely to be recorded only 'in passing'. For example, it is through obituaries that help given by Sir Stanford Downing at the Ecclesiastical Commissioners (J.S., *SWHEM Quarterly Bulletin*, October 1933b) and Sir George Duckworth at the Ministry of Munitions (*SHHM, Quarterly Bulletin*, July 1934c) comes to light. The role of Parker Morris in helping the Society was better known. As a prominent town clerk at the time he used his contacts to further the work of women managers as well as giving them legal advice from time to time. The fact that the Society contained many middle- and some upper-class women probably helped with these contacts.

OVERSEAS CONTACTS

Octavia Hill had already in her day been recognised internationally for her work in housing management and had made many contacts overseas, especially in the USA and Holland (Hill, 1956: 184, 185). These contacts were continued and extended by the Society, both in receiving visitors and in individual members' or official visits abroad and in publishing reports in the journal. For example there was considerable correspondence with the USA.

In 1934, the National Association of Housing Officials, Chicago, requested Sir Raymond Unwin, P.P.R.I.B.A., to lead a team of lecturers on housing matters to the United States of America in the early Autumn of 1934, in connection with the development of President Roosevelt's housing programme.

(SWHEM, 1934b)

Miss Samuel (a member of the Executive of SWHEM and the person who had pioneered the work in Bebington) accepted an invitation to take part in this tour (*SWHEM Quarterly Bulletin*, July 1935).

Another example of contact over a period of time was South Africa. In January 1934, an unofficial representative of the Cape Town municipal authority (Mr Hankey) asked the Society to draw up a list of questions which they would like answered by the Cape Town municipal authority before considering the appointment of a manager (SWHEM minutes, 14.4.34). Preliminary negotiations being satisfactorily completed, Miss Margaret Hurst, later Mrs Margaret ('Peggy') Hill, was appointed

to assist in Housing Administration under Cape Town City Council, her especial work being the development of the Octavia Hill system in Cape Town and the training on these lines of European and non-European women.

(SWHEM, 1934b)

Apparently it had been the Women's Municipal Association in Cape Town who had first suggested the appointment of a woman housing manager to the Municipality (*The Times*, 1934). The appointment of a young (aged 26) Englishwoman to manage property in Cape Town did not pass without comment in the local press (*Cape Argus*, 1935), and there were some doubts about her capability to deal with such a difficult situation. In fact it proved possible to establish contact with the tenants by regular visiting (though a woman could not collect rents), to interview applicants and to visit acquired slum property, to begin to train South African women for the work of housing management (Hurst, 1937), and to carry out considerable publicity for Octavia Hill work (for example, *Cape Argus*, 1935). Contact with South Africa continued after she returned.

Visitors from a number of other countries were in touch with the Society from time to time, for example Australia, Czechoslovakia, Palestine and Sweden (SWHM, 1939), Germany (*SWHEM Quarterly Bulletin*, October 1933a), France (*SWHEM Quarterly Bulletin*, July 1934d) and Denmark (*SWHEM Quarterly Bulletin*, July 1934e). The

Society was particularly interested in Holland, where work had been started after an early contact with Octavia Hill (Hill, 1956). Many Society members disagreed with what was known as the 'Dutch System' of separating the really bad tenants into a 'control' under strict supervision which included 'frequent inspections of the house and compulsory weekly baths for all inmates' ('H.R.T.', 1937). These contacts provided a broader view of housing practice and added to the Society's prestige, but only South Africa provided employment opportunities.

It is clear that in the 1930s the Society played a vital part in expanding women's employment in housing. In face of the wide-spread discrimination against women described in Chapter 4 it pro-vided a support base. Because its training and recruitment were closely controlled, the Society's members formed a fairly coherent group with a clear idea of professional identity.

While many of its activities were usual for an organisation of that type, the Society pursued publicity and contacts with employers particularly energetically. The various publications helped to spread information about the women housing managers, encourage recruit-ment and prepare the ground for employment opportunities. Contacts with other women's organisations helped to share knowledge and gain support both for specific housing policies and for the employ-ment of women managers. Contact with other housing and local government bodies provided further publicity and chances to meet potential employers. Public speaking to a broad range of organisa-tions fulfilled similar purposes and the overseas contacts brought some possibilities of employment, knowledge of new developments and increased prestige.

The work that the Society undertook with individual employers was obviously vital in translating tentative enquiries into actual employment. The stand the Society took on conditions of employ-ment and salaries would help to stiffen the resolve of women who, knowing they were in a weak position, might have accepted posts with lower salaries or succumbed to the 'welfare' role.

In all these ways the Society was an important element in encouraging the employment of women housing managers. But in doing so it often used an argument of 'special aptitude' of women for the work, which implicitly reinforced stereotyped attitudes to the capabilities of women. In addition, some of the Society's advantages could be turned to disadvantages. The fact that the Society was wedded to a particular form of management which could be seen as paternalistic meant that they were vulnerable to attack from the Left.

The form of training which socialised students so well was also a block to any substantial expansion in numbers and limited access for working-class students. Even the existence of an all-woman Society which could campaign so well for the employment of women meant that the men employed in housing, including the most able and the heads of the largest organisations, automatically had had to join a different organisation with which there was little contact. However, it seems fair, after an examination of the difficulties which women faced in housing employment in the 1930s, to conclude that it is unlikely that women would have made as much progress, especially in local authority employment, without the energetic work of the Society.

6 The war years

Most histories of housing pay surprisingly little attention to the war years. Burnett's social history (1978), for example, moves straight from describing housing in 1939 to trends from 1945 onwards with hardly a mention of the war. Merrett (1979) gives it a little attention, but Berry (1974) devotes just half a page to its effects. It is relevant to this study to ask what happened to the administration of housing during the war? And how did this affect the Society of Women Housing Managers and its members?

PREPARATIONS FOR THE WAR

Prior to the outbreak of the war preparations were being made but the planning for housing was quite inadequate. Central government was preoccupied with expectations of panic among the civilian population and ignored evidence about the damage that could be caused by bombing. The Treasury was reluctant to commit money to the preparations and local authorities were reluctant to act without money.

The country was divided into three zones for evacuation purposes – evacuation, care and reception – but these had to be changed from time to time with the fluctuation of the war fronts. The story of evacuation has been told elsewhere (Ministry of Health, 1940a, 1941a; Women's Group on Public Welfare, 1943; Calder, 1969).

The need to give practical help to those made homeless by enemy action was partially recognised and the local authorities were to be responsible, helped by central finance. The original plans were for emergency stations providing temporary shelter with few facilities, assuming that those bombed-out would quickly move on to be rehoused by friends and relatives. The Blitz of 1940 revealed that these plans, and their implementation by the local authorities, were utterly inadequate.

THE BLITZ AND AFTER

The extent of destruction of housing by bombing had not been foreseen by the government (2,250,000 people in the UK were made homeless by the raids of 1940–41). London, which suffered especially badly, had even more administrative problems than other areas because of the multiplicity of local authorities. Lack of preparation resulted in appalling conditions in overcrowded and under-provided rest centres and shelters. People helped themselves by opening up tube stations as temporary shelters or 'trekking' out of cities to sleep anywhere they could find in the country (Harrison, 1978: Chapters 5, 6 and 7). Voluntary evacuation was aided by the government, but could increase the overcrowding of certain areas.

The problem of housing war workers began to get more acute as war factories, which had been relocated in the west and north-west of England and in Wales, came on stream and recruited workers. The Housing (War Requirements) Committee was set up in 1940 to co-ordinate demands of different departments but

> [The Committee] is too much under the control of the Ministry of Health who insist on going through their lengthy procedure for getting to know the available lodging accommodation *after* we have learned by experience that none exists; who thereafter pander to hopelessly dilatory and evasive local authorities whose only object in life is to extract subsidies from the state and who don't mind whether the houses are finished before the war or after.
>
> (Dowall, 1940)

The government rapidly pushed ahead with the improvement of rest centres but the next problem was where to rehouse those who sought refuge in them.

> 'And then, the idea was that we must get people out of the rest centres, and this was so typical of the government, they said people just mustn't stay in the rest centres for more than four days . . . and you must give them priority. Well, of course, if you gave them priority, everybody went to a rest centre in order to get housed, so we said very soon we must give everybody the same priority and you must tell anyone if you go away and stay with your friends. We will still house you, and then of course a lot of people did and they got a billeting allowance by going to friends.
>
> (Interviewee, London housing manager)

H. U. Willink was appointed as a special commissioner to deal with London's homeless and a general reorganisation and clarification of responsibilities began.[1] The functions of the metropolitan boroughs were defined as: billeting, requisitioning, the salvage and supply of furniture, and 'welfare'. The boroughs were told to appoint an executive rehousing officer to supervise the work under these four heads and to furnish the officer with sufficient full-time assistants. All expenditure would be reimbursed in full (Titmuss, 1950: 288).

The Society of Women Housing Managers had been in touch with H. U. Willink (SWHM minutes 5.11.40). This seems to have had some effect since the minutes later record that four enquiries about the appointment of rehousing officers (Camberwell, Greenwich, Hampstead, Lewisham) had resulted from that interview (SWHM minutes 16.11.40). The work was extended when H. U. Willink began to appoint trained social workers (Titmuss, 1950: 289). An interviewee who gained an appointment to a London borough where rehousing had been in some degree of chaos said:

'the heavy bombing had begun and the rest centres were getting full . . . and this friend of mine . . . she'd been involved with a lot of housing . . . she always says that she started off my job in a sense because she was one of what they called Willink's Inspectors . . . he had "Willink's Young Ladies", they were called . . . and they were trying . . . because the boroughs were in an awful mess, so she told [the London borough] "You'd better have trained housing managers." So they advertised the job' [and the interviewee got it].

The kind of amateur administration the boroughs had been practising was described by another woman manager appointed at this time:

'I think the Ministry had said, you know, that you'd better get someone who knows something about housing. . . . The work had been done by the librarian . . . assessing rents and requisitioning properties . . . the first thing the treasurer wanted was the arrears down (but the rent records were in such a state).'

(Interviewee, London housing manager)

Requisitioning

Requisitioning of unoccupied dwellings had been introduced in 1939. It posed administrative and management problems but these were gradually eased with ministerial advice and shared practice. In

1943 local authorities were given power to requisition properties for the inadequately housed as distinct from the bombed-out homeless (Ministry of Health, 1943c). Though the numbers of requisitioned houses were not huge they made a critical contribution to relieving stress.

> 'Day began with finding out how many people were in the rest centres – then requisitioning and making big houses into separate places if one could . . . flung in kitchens where we could. . . . Requisitioned a rather posh block of flats and two families would share one six-bedroom flat – put two cookers in. People were marvellous.'

Relationship between central and local government

The initial weakness of services for the homeless and the need for government intervention produced a much closer relationship between central and local government officers than had been usual in the past. Looking back, Miss Thrupp, the manager of Mitcham, who received an OBE for her services during the war, described the arrangements for liaison between the Ministry and the local authority.

> Many housing managers became billeting and rehousing officers and, probably for the first time in their careers, came into direct contact with civil servants acting for the appropriate Ministry. This was an interesting development because, on the face of it . . . it might appear that the local councils were, so to speak, cut out of the picture. . . . Mitcham was one of a group of contiguous boroughs and a civil servant, attached to the Ministry of Health, was the group rehousing officer.
>   To him all problems relating to the circulars connected with 'the bombed out' could be referred. During heavy bombing he called once a week on each authority, besides always being available on the telephone (when it was in order) both at his office and at home . . . he was always present at monthly meetings of the group . . . and . . . suggestions were put forward through the group rehousing officer to the Ministry concerned. In my opinion, this contact between the central government and the local authority was a complete success . . . but I do emphasise that this was an instance of removal from the local authority of responsibility for even the administration of the central government's policy.

> (Thrupp, 1947)

The group rehousing officers provided detailed reports to the Ministry of the state of rehousing provision in their areas (for example, Ministry of Health, 1941b). Miss Thrupp's account describes London region practice; contact may not have been as close elsewhere.

It seems that in a number of local authorities this new rehousing function was kept rigidly separate from the management of the normal council stock In others, however, it was combined (interviewees). In reception areas 'the billeting of the homeless and of evacuees constitutes a single problem and is dealt with by the same organisation' (Ministry of Health, 1942a).

And in the various villages it was Mr Jones the school who was usually the billeting officer. And I remember going to see one, who was conducting about four different classes in a large hall, and cooking his own sausages on a donkey stove in the middle of the thing at the same time and interviewing me.'

(Woman housing manager working in rural area)

CONSTRUCTION AND REPAIR

There was no government department with responsibility for co-ordinating different building programmes. It was only in October 1940 that a proper system of licences for private building and a department to operate it was established (Hancock and Gowing, 1949: 174–5). Effective co-ordination of housing construction or repair was never achieved so there was a series of crises as government building expanded, houses were destroyed or damaged and materials or labour ran short. At an early stage the duty of supervising and carrying out repairs to housing accommodation rendered unfit for human habitation by air bombardment or other war action had been given to the housing authorities. They also had the duty to compile a return to the District Valuer on the number of buildings which were destroyed and damaged (Ministry of Health, 1939).

Titmuss (1950: 277) comments on the unreliability of the war damage statistics which are based on these returns. In 1940 local authorities were very short of building labour and the schemes were administratively cumbersome. The first crisis occurred after the Blitz. It was clear that the repair of damaged houses had to be given increased priority. The government created a special organisation to assist local authorities. Mobile squads were set up, including men specially released from the army, and these could be switched to any heavily attacked area. (During April 1941, the Directorate of

Emergency Works put 16,000 repairers into London in little more than a week.)

> By August 1941 over 1,100,000 houses had been made wind and weather proof and so just habitable. . . .With boarded windows and tarpaulins on the roof, home might not be sweet, but at least it was home.
>
> (Calder, 1969: 192)

By 1943 the housing crisis was spreading and the local authorities complained:

> The type of labour at present available involves the employment of men over 60 years of age and boys. As a result, progress is slow and actual labour costs are proving to be far in excess of estimated costs.
>
> (Parker Morris, 1943)

Government began to give a little more support for housing by easing labour restrictions (Ministry of Health, 1943b). In 1944 the government decided to proceed with the immediate production of temporary housing, making extensive use of prefabrication (Merrett, 1979: 237). By the time war ended housing problems had multiplied to a daunting extent (Titmuss, 1950: 430).

THE SOCIETY AND WOMEN'S EMPLOYMENT DURING THE WAR

Control of women's employment

By 1941 the plans for enhancing war production began to result in large demands for labour and measures were taken to bring about an increase in the number of women workers by introducing registration of women, with some exemptions. As time went on the 'comb out' of women from non-essential jobs became more severe and after December 1941 women were liable to be called up for service in the auxiliary forces and civil defence. Married women and women with children under 14 were exempt. These regulations were brought in for younger women and gradually extended to older women (ILO, 1946: 16–22).

With legislation increasingly affecting the employment of women, it was necessary for the Society to take steps to secure the status of its members. By September 1942 the Society had secured reserved occupation status for its members. The housing management course was shortened to a year to make more effective use of students in wartime conditions (*SWHM Bulletin*, January 1943: 2).

Negotiations with employers

As in the 1930s, the Society sought opportunities to publicise the work of women managers, to follow up enquiries by personal contact if possible, and was even prepared to negotiate conditions of work and salaries. Initially the war had not created any extra demand for housing managers (SWHM minutes, 30.9.39).

> The likelihood of new openings for women caused by the calling up of male members of staffs for war service was suggested, but so far most local authorities were more inclined to cut down on staff for the sake of economy than to make new appointments. Miss Upcott reminded members of the danger of taking on men's jobs in circumstances which would not allow the full application of our principles of management.
>
> (*SWHM Quarterly Bulletin*, January 1940: 4)

Miss Upcott had her own reasons for reservations about war employment as she had been one of the managers dismissed from the Ministry of Munitions after the First World War. Miss Larke, on the other hand, considered it advisable for us to take on work of a temporary nature wherever possible, as it would be bound to give some impression of our capabilities (*SWHM Quarterly Bulletin*, January 1940: 4).

During the First World War, Miss Larke had followed this course. The likelihood of new openings soon became a reality as the war continued and the billeting and rehousing work described earlier expanded.

> Many enquiries have been received during the past six months from local authorities and other landlords wishing to employ women to replace members of their housing staffs who have been called up for military service. In some cases . . . the conditions have not been sufficiently favourable for Octavia Hill work for the Council to feel justified in advising members of the Society to apply for the posts. Since the Blitzkrieg began there have also been enquiries about the employment of members as rehousing or billeting officers in bombed areas; some members are combining these duties with their usual work.
>
> (*SWHM Quarterly Bulletin*, January 1941: 1)

Opposition to the appointment of women or the Octavia Hill system was still found and could be quite open.

> *Lincoln.* An advertisement had appeared for a senior housing assistant, applications being invited 'from males only' . . . the staff at Lincoln had written asking for advice and SWHM Council agreed 'that a letter be sent to the Town Clerk expressing regret at the situation and pointing out that the Lincoln decision placed the Octavia Hill assistants and the students in a difficult position'.
>
> (SWHM minutes, 17.6.39)

This action did not have any effect, however. It seems that a man was appointed and the students encouraged to join the Institute of Housing. The Society continued to refuse to recommend or advertise appointments where it felt that conditions were unsatisfactory from the beginning, either because of the nature of the work or the mix of duties. For example:

> *Huddersfield.* An enquiry had been received through Miss F. L. Sawers for a woman manager mainly to undertake 'social uplift' among tenants. Agreed that such compromises could not be accepted, especially in view of the present demand for trained workers.
>
> (SWHM minutes, 16.3.41)

The Society did, however, accept appointments in Lambeth which were temporary, 'for the sake both of the men they would replace and of their own status' (SWHM minutes, 8.6.40). But this arrangement did not work out very well.

Central government appointments

Ministry of Supply

As we have seen, the need to accommodate workers for factories concerned with urgent war production began to clash with other accommodation needs. Manufacture had been spread out as far as possible to rural areas. In some areas, estates had to be be built and managed by the Ministry concerned because of lack of other accommodation (Hill, 1981).

The experience of the First World War housing for munitions workers and the involvement of women managers was remembered in some places.

Housing of Munitions Workers. The National Council of Women, wishing to avoid such waste of material and public funds as was caused by the type of temporary Housing Estates erected during the last war, urges the Government to provide an adequate and efficiently administered system of Housing Estates for workers in the new munition and other factories.

(SWHM minutes, 8.6.40)

The July 1941 minutes record that Miss Larke (who had worked during the First World War) had written to suggest approaching the Ministry of Supply with regard to the management of houses being built near the new factories. But it was agreed that no action be taken in view of the scarcity of qualified workers (SWHM minutes, 13.7.41). But the wheels were turning and 'somebody remembered'. The Ministry of Supply advertised for a housing manager and Margaret Hurst applied for and got the job.[2] She was appointed as a housing manager but in a few weeks became 'Administrator for Married Quarters' (Hill, 1981).

We have already seen that the Society had contact with H. U. Willink over the appointments to some of the new posts in London boroughs after the Blitz. It also seems to have been closely involved with the Housing Management Adviser's post at the Ministry of Health and other central government appointments.

Ministry of Health. The chairman reported on an interview she and Miss Tabor had had with Miss Aves, Senior Welfare Officer, on 29.7.43, and also on an interview she, Miss Hurst and Mrs Cockburn had had with Miss Russell Smith with reference to a proposed appointment, since sanctioned, of a member of the Society to act as a liaison officer to local authorities in connection with questions of housing management. Miss Alford, Miss Hurst and Miss Thompson were being considered for this appointment.

(SWHM minutes, 19.9.43)

THE SOCIETY – WARTIME ORGANISATION AND ACTIVITIES

The administration of the Society

By September 1939 a decision had been made by the Housing Centre to close, in view of the outbreak of war, and the Society's work had to be carried on temporarily from the secretary's address in London (SWHM Circular letter, 4.9.39). The office therefore moved to

Oxford with the secretary, Dr Miller (SWHM minutes, 30.9.39), but this proved inconvenient and by July 1941 the London office was re-opened with a part-time secretary, Mrs Cockburn, having special responsibility for students while the acting secretary remained mainly in Oxford (SWHM minutes, 13.7.41). When Dr Miller was due to leave Oxford, the office was finally moved back to London in August 1942 (SWHM minutes, 12.7.42).

Wartime arrangements also involved the setting up of an Emergency Executive (SWHM minutes, 9.9.39) though it was resolved that the Council should meet every three months and that the work of the Society should be maintained as far as possible. Despite wartime conditions, efforts were made to continue the support which the Society had given to its members. AGMs were held each year and, apart from 1940, these included visits to property and talks, as well as the business meeting.

A conference took place each year, but in 1942, 1943 and 1945 this was held in London because of the difficulties of wartime travelling and accommodation. However, the Society managed to use a provincial location in some years: Birmingham in 1940, Lancaster in 1941 and Hereford in 1944 (SWHM Annual Reports, 1940–45). A London regional group of managers formed and this also met to discuss wartime problems; on some occasions, those managers concerned with rehousing also met specially (SWHM, 1941: 5). Managers who had been working during the war commented how essential the support of the Society and of other members had been in hearing problems and keeping up morale. The Junior Organisation also became an important social focus for younger members. A number of meetings were held in connection with reconstruction. Society members continued to address meetings of other organisations as part of the general propaganda but details of these meetings were no longer given in the reports (SWHM, 1943a: 7).

The bulletin was reduced to a wartime paper saving format, losing its card cover and reducing print size in May 1942. Sometimes there were longer intervals between publications, for example between January and June 1944. But on the whole the content and style of the bulletin was maintained.

Membership

As Table 6.1 shows, the war period was one of growth for the Society. Almost a hundred more ordinary members were added over this period and the number of students had doubled by 1945.

Table 6.1 Membership of the Scoiety during the period 1938–45

	31 Aug. 1938	*31 Aug. 1945*
Fellows	26	35
Ordinary members	159	256
Total	185	291
Honorary members	30	27
Associates	63	79
Students	60	122

Source: SWHM Annual Reports, 1938–39 to 1944–45

Training

The Society continued to operate its training scheme through the war, though adjustments had to be made. Exemption from conscription was gained for students. There was some criticism of the training offices by students, though no details are given.

Possibly a sign of changing times was the fact that the difficulty of financing students began to be discussed. A letter to the bulletin in May 1942 raised the whole question of recruitment to housing management and suggested, especially in view of the likely demand for post-war housing reconstruction, that there should be a long-term policy. Comparisons were made between the publicly financed flow of recruits into teaching and the difficulties of the would-be housing manager:

> We are familiar with the difficulties which face the prospective student for Octavia Hill work. She is told that coming straight from school, she is too young and inexperienced, and that she must fill in a year or two with some other training. She must face the fact that for several years she will be unable to support herself and, unless her family are fairly well off, she will probably decide to do something else. In this way the Society loses many promising students, and suffers, perhaps, in that so large a proportion of its members are drawn from the 'comfortable' middle class.
>
> We should like to suggest, therefore, that a joint scheme might be worked out to provide a grant-aided avenue for social and welfare workers, similar to that already in existence for teachers.
>
> (Members at Newcastle under Lyme, 1942)

Though some financial help was given by the Society this was very limited and it was difficult for students from less well off backgrounds to qualify, though the gradual growth of local authority trainee posts could help (interviewees). Thus the class basis of the Society would remain limited until better help was available.

At the annual Provincial Conference at Hereford in May 1944, there was a more thorough discussion of training, initiated by the chairman of the Junior Organisation. Some of the dissatisfactions of students were aired. These included the difficulty of correspondence study after a day's work, the constraints of correspondence tuition and some criticism of the syllabus (Snook, 1944).

A more radical departure was the recognition of the need to do something for older staff working in SWHM offices who would not be able to undergo the rigorous training which the Society normally required. This was announced in the Annual Report of 1944–45:

> Scheme for training Uncertificated Assistants: . . . At an Extra-ordinary General Meeting held on July 8th at Walston House, a training scheme for uncertificated assistants was approved and a resolution passed empowering the Council to set in motion the necessary procedure for the alteration of the Article of Association to admit a new class, to be known as Licentiates.
>
> (SWHM, 1945: 3)

The Junior Organisation

The Junior Organisation continued to flourish throughout the war years, though wartime circumstances could make it difficult. The Midlands Group started a newsletter to keep offices in touch with each other and the North-West Group held several informal meetings. The London Group had to abandon its original programme but decided, until evening meetings were possible, to hold Saturday lunch-hour meetings. 'That many members attended was proof of their keenness' (SWHM, 1941: 5,6). Interviewees who had been involved with the Junior Organisation during the war spoke warmly of the support it had provided.

The admission of men

This issue, which was to become much more prominent in the post-war period, had already surfaced during the war. At the AGM in November 1943, the motion was put forward

> That in view of the immense development of housing schemes likely to take place after the war and the public recognition of the value of trained management, the Council be asked to consider

making the Society's training available to men as well as to women.

The motion was proposed by Jean Thompson, manager at Southall, and seconded by Miss Alford, both members already working in local authorities. Jean Thompson said:

> A number of men would be doing housing work similar to ours after the war; was there to be no alternative training for them to that given by the Institute of Housing, whose principles were opposed to ours? . . . If we insisted on one-sex training we might hold up a big advance.
>
> (*SWHM Bulletin*, January 1944: 4)

The main arguments put forward against the resolution were as follows:

1 The limited capacity of offices for training; that what places there are should be kept for women.
2 The danger of swamping the Society by admitting large numbers of men.
3 The likelihood that the men who applied for training would be of an inferior type.
4 The difficulty of inevitable differences in salary.
5 That men would get the best posts and leave the routine work to women.
6 That men would not get the entry to houses as women can.
7 That the Society has made its way as a women's society and its members have done their work as women. It is not exclusive but a speciality.
8 That this is a bad time to consider such a resolution because:
  (a) the Society is not large enough nor public opinion enlightened enough for it to be an established practice to appoint women to managerial posts.
  (b) the economic position of women after this war will be as bad as after the last, and we should not give up anything we have won.

> (*SWHM Bulletin*, January 1944: 4)

The discussion carried on in the bulletin was revealing of attitudes and stereotypes and will be more fully considered in the next chapter. The fact that the issue had already surfaced during the war period is indicative that some members had already realised that SWHM's position would be weak when housing employment started to expand again.

THE WORK OF WOMEN MANAGERS DURING THE WAR

Local government

Keeping the service running

The lack of preparation and the critical conditions of the Blitz period, described earlier in this chapter, meant that housing staff had to work long hours, often in very poor conditions:

> 'We started by having our office combined with the public assistance office because they had to get ration books and money and all those things . . . if the house really was demolished then of course they had to have help with clothing and identity cards and ration books and all that, so they had a long journey to do through all the officials ending, probably, with us – Where were they going to live? Where were they going tonight?' . . .
>
> 'And then we had a bomb on the power station and that put all the electricity . . . all the lights off . . . and we were working in an underground shelter, you can imagine what the temperature was like, and the air. And because, of course, we had to send out and buy up all the candles . . . the Council hadn't got enough and . . . they brought us all the street lanterns in . . . well you can imagine how that heated us up . . . 500 people all night, and all day you see they were waiting to see us. . . . We flooded, because of the sewers, and the pumps had all gone.'

Naturally there was a lack of suitable staff.

> 'We had a very mixed staff, it was fascinating really. We had a few of the local government people . . . there was a chap who had a gammy leg in the library so he wasn't in the services. We had women who'd never worked for pay at all or who'd done sort of COS, done curious jobs – married women who hadn't perhaps worked before they married, who were called billeting officers, but in fact they had to go out and look at property.'
>
> (Interviewee, London housing manager)

Offers of help from other Society members were very welcome:

Of course we worked seven days a week . . . we used to take Sunday off and we were called back if there was a raid . . . well now, other members of the Society wrote or rang us and said they'd like to come and help us at weekends, and they came, two or three people came, and this was wonderful . . . and this impressed the Council so much, they couldn't get over it.

'Staff rushed to London to help in the Blitz – had three to four qualified staff, the rest were people like acquisitions and billeting clerks. The young wanted to be "in it" and come to London and do a dangerous job.'

(ibid.)

The job of rehousing

Managers were working under considerable pressure, with the threat that accommodation in the rest centres would block up completely or that the supply of rehousing would dry up. In addition they had to deal sensitively with people who were often suffering from shock, but still had their own preferences as to where to live.

'There'd been bad bombing in September [1940] . . . and then there was a very bad patch in the spring. And I remember the awful fear was that the rest centres would be full and there'd be nowhere for them to go . . .'

'And the thing we were always threatened with was that billeting would become compulsory. They couldn't get people out of the rest centres . . . they would make billeting compulsory and we would have to compulsory-billet people. This was a sort of nightmare . . .'

'Some of them came in having literally been bombed out that morning. I mean, some . . . the first thing they wanted to do was to get in somewhere. . . . Others . . . couldn't think about it for days. This is what we learnt at the rest centres: that for some people you just had to go in and talk to them and not try to offer them anything, just let them . . . just make them feel that you are on their side. Because . . . if they thought that you were pushing them out, it was just like they thought you were pushing them out of their slums.'

Four of the interviewees had similar experiences with rehousing:

'We never went dry [i.e. ran out of housing], we'd come right down to the last two and not know what on earth we were going to do the next day . . . but it was a real widow's cruse . . . we never went dry. We had to do our duty there, one night a week we had to be on duty all night . . . we used to go out on to the site and that was a tremendous help to us, because the Sanitary Inspector's plans of course had to be detailed, and they couldn't get those done and duplicated and out to us until perhaps midday. So we would go and take note . . . of the houses. We might say, well we can rehouse beyond number 22 or . . . We never housed on the spot . . . they must go to the rest centre, that's what it's there for.'

'[Rehousing] was in some ways rather like slum clearance, this was what was so fascinating. . . . I mean, I know when I went I thought it would be a sort of not exactly a military operation, but you know, somebody would be bombed out and they'd be housed . . . but instead of that, what people were worried about was . . . what they were doing with their dog and where they were going to keep the pram.'

Suburban and rural areas

Even in areas identified as neutral life could quickly develop hazards.

Romford is a neutral area for purposes of evacuation, although the German Luftwaffe has other ideas as to its desirability as a target but, despite the numerous bombings and the frequent craters to be met with in the vicinity, the estates have escaped material damage, and the tenants, generally speaking, are in good heart . . .

The proximity of the estates to the areas most affected by air raids has led to numerous applications for accommodation by people who have been bombed out of their homes. Their needs are of course always immediate. If I am unable to offer them accommodation on the spot they rarely return to me, and the number of refugees rehoused on the estates is comparatively low.

(Carey Penny, 1941)

In one rural authority, an interviewee recalled, the town clerk

'wanted to have a housing department, they'd never had one before. This was still in the war you see, and you had to be billeting officer as well because . . . in a curious way [this] was both a reception area and had two or three aerodromes so . . . there was a lot of pressure on the accommodation in the area.

'And I remember I was the only woman interviewed and I never knew how they'd took to me. There were four men who were collecting under me, and I don't think they really minded too badly . . .

'So the main thing was getting the waiting list in order . . . so that people had a chance even if they weren't known to the local councillors. . . . I don't think I would have been there except for this clerk . . . who was very active and wanted to clear up all kinds of scandals . . .'

Miss Hort, working in another rural authority (Wokingham) similarly struggled with some long-standing and some new problems:

My biggest problem is repairs, and it may well prove a despairing one while war conditions last. The Addison houses have their usual afflictions of dry-rotted floors and penetrating dampness. Inside decorations have been done in a most haphazard way and, as soon as I appear, the tenants fall on me and ask to have rooms done up which have not been touched in some cases for ten years. The Surveyor, who was also the Housing Manager, was called up at the beginning of the war and, since then, first the Engineer and then the Sanitary Inspector have carried out the repairs, but without any real system. . . . Very few of the houses are on main drainage, and I am learning the full horrors of cess-pools. Most of the houses have earth-closets, but in Twyford there are W.C.s and the cess-pools are in the front gardens, so that when they overflow, as not infrequently happens, they run down into the street . . .

The Housing Committee has been suspended during the war, but the Chairman, who is a woman, is an able and energetic person . . . When a vacancy occurs in a Council house, the practice is for the new tenant to be chosen by the Chairman and the Councillor representing the parish concerned, who is supposed to have an intimate knowledge of all the local applicants. I do not yet know what degree of impartiality exists among the Councillors, but I foresee that this practice may be open to objections.

(Hort, 1941)

Working for national government

Ministry of Supply

By January 1943 there were twelve housing managers at the Ministry of Supply (*SWHM Bulletin*, January 1943). 'All Octavia Hill trained . . . this was Ministry policy' (Hill, 1981). There were regular meetings in London to keep these managers in touch and provide them with support as they were often working in very isolated roles (interviewees):

> 'This was one of the good things about the Ministry of Supply. You went to a factory and they said "Ooh, you know about housing, thank goodness, here's the files, here you are, anything you want just ask us, goodbye." You knew you were really wanted; it wasn't a case of your wanting to do the job, it was a case of their wanting you to do the job. And the whole attitude of the Civil Service was good in that way. They had fought the battles before; you were accepted on your merits. I think it was a contrast to local government, where you had to convince them all the time you could do it, until they got used to you.'

Despite the difficult, isolated and sometimes exhausting work, many of the managers valued this work experience. A number of them went on to important jobs, such as posts with the first New Towns, where they felt that the wartime experience was regarded as significant.

Ministry of Aircraft Production and the Admiralty

The Ministry of Aircraft Production usually got local authorities to manage their houses. 'Generally, the MAP estates are managed by local authorities but, where this is not, or has not been, practicable, independent agents have been, or would be, appointed' (Ministry of Aircraft Production, 1941). After a time women housing managers were appointed but it was only a small establishment. One interviewee was appointed to manage an estate which the local authority had felt it could not cope with. 'The official title was Ministry of Aircraft Production Housing Manager . . . for about 500–600 temporary bungalows, prefabs with grey walls, bricked but unplastered inside. Locally called the piggeries'.

As with MOS work, the estates were occupied by people of mixed social class. This housing manager, however, stuck to her own principles of fair treatment.

> 'The factory used to send down batches of names of people coming and which wife they were bringing or whether they were going to start with a new one. . . . For example, the Managing Director already had his wife, or somebody called Mrs X, in one bungalow; then he brought another one down and installed her in another bungalow. I told him he couldn't do it; after all a person's life is their own but he couldn't have two bungalows.'

(Wartime MAP Manager)

Society records show that contracts with government departments were still further extended by the appointment of two members as Officers for Housing Duties to the Admiralty (SWHM, 1944: 4).

Ministry of Health

An appointment which seemed important at the time, both for the Society and for housing management, was that of an adviser on housing management to the Ministry of Health. Unfortunately, no information about this has been found in Public Record Office files and apparently most internal records have been destroyed. A little information is available from a subsequent holder of the post (Fox, 1974) and from the interview with the first post-holder, Margaret Hill:

> 'At the end of 1943, Willink was at the Ministry of Health and was very keen on the welfare side and wanted to stress housing management at the Ministry and so they advertised for an adviser . . . and again I was lucky to get it . . . and so for a short time I was adviser on housing management which was a jolly difficult job because my job was to go round to all the local authorities . . .'
>
> (Hill, 1981)

It is likely that Margaret Hill's experience at the Ministry of Supply in charge of a team also seemed good background for this job.

After Margaret Hill left the Ministry in 1945 (SWHM, 1945) she was succeeded by M. Empson, who was also a member of the Society and who remained in post for 22 years (Fox, 1974). The establishment did not expand in the way originally intended.

> 'We were meant to have regional offices and have the whole thing going . . . very much for housing management but local authorities were pretty individualistic, as you know, and I think it's all developed in different ways.'
>
> (Hill, 1981)

So far as the wartime period is concerned, however, the appointment of a member of the Society as the sole adviser on housing management would seem to be a signal recognition of the work and standards of Society members and one which would be expected to lead to even greater standing for its work.

Work on reconstruction

From a fairly early stage in the war discussions about post-war housing began and the Society was involved in meetings and committees on this. One member felt that there were possibly too many meetings about this and 'they were rather a waste of time' (interviewee).

For example, the National Council of Women set up a sectional committee on post-war reconstruction on which the Society had several representatives, though representation from this group was later withdrawn because the meetings were held at an inconvenient time for housing managers (SWHM minutes, 12.7.42). The first of several meetings organised by the Society on planning was held at the Housing Centre in September 1941. The decision was made to form groups within the Society to look at different aspects of planning and Professor Holford was invited to speak on Reconstruction and a National Plan at the Housing Centre in October (Hamilton, 1941).

The Society was asked to provide information for the Nuffield College Reconstruction Survey (Strange, 1941). Later the planning groups became involved in a questionnaire (SWHM minutes, 15.3.42) and the work on this was incorporated in the Society's evidence to the Dudley Committee.[3]

A Women's Conference on Planning and Housing was organised in May 1942 at the RIBA by the Housing Centre with the Women's Institutes, the Women's Electrical Association, the National Council of Women, the Women's Co-operative Guild, the Society of Women Housing Managers, and others. It was addressed by Professor Patrick Abercrombie on national planning, and Henry Willink chaired a Housing Brains Trust. At the end of this conference Lady E. Simon moved a resolution that 'Women were not adequately represented on the various committees set up to deal with post-war housing and planning' and urged the government to appoint a committee of women qualified to advise on these problems. She pointed out that, during the war, vacancies on local councils were filled by co-option and this was a chance to get women on. Women should talk about housing and get it discussed while opinions were still fluid (Hort, 1942: 6).

As we saw in Chapter 4, efforts had been made earlier to get better representation of women on CHAC – this effort was little more successful than previous ones had been.

The Dudley Committee noted that 'the Minister was referring to the main committee the question of women's representation in regard to sub-committee's work' (CHAC, 1942). Later Miss Megan Lloyd George, MP joined the sub-committee's membership though it is not clear whether this had anything to do with the representations or not. The SWHM minutes also note that Miss Crewe was representing SWHM on the *ad hoc* committee which the Women's Advisory Housing Council had been asked to set up to help the sub-committee sift the evidence submitted (SWHM minutes, 10.5.42). The Dudley Committee minutes record the attendance of this *ad hoc* committee in April 1943 (CHAC, 1943a).

The Society had asked to defer its evidence to the Dudley Committee until the results of the questionnaire were ready (SWHM minutes 10.5.42). A report based on the questionnaires was published in May 1943 and the bulletin records that evidence based on the results of the planning questionnaires was submitted to the Sub-committee on the Design of Dwellings of the Ministry of Health Central Housing Advisory Committee, which also received oral evidence from members of the Society (*SWHM Bulletin*, June 1943: 2). The evidence submitted, in a booklet of 52 pages, was based on a questionnaire undertaken by 35 managers and including 2,077 tenants selected on a sample basis to give representation to different income groups and family types (SWHM, 1943b). It therefore represents a substantial effort for the Society. It remains uncertain how far SWHM's painfully gathered evidence was used by the Dudley Committee.

The Society's report was quite wide in scope, stressing the need for a variety of dwellings to cater for the fact that there was a variety of taste and expectation. It made a large number of specific recommendations, including the need for playgrounds both for cottage estates and flats, sound-proofing for houses and flats, but 'for families, flats are to be avoided wherever possible'. The report recommended that each block of flats should provide its own room for tenants' meetings and its own communal workshop. It also dealt with housing for the aged, housing for the large family and the vexed question of the problem family, where it stated that 'segregation of the problem family is to be avoided; so also is specially designed accommodation with cheap or inferior equipment' (SWHM, 1943b).

The Society was also represented on a number of other committees concerned with post-war housing. In 1943, for example, there were Society representatives on the CHAC Sub-committee on Rural Housing, the Ministry of Works and Planning Directorate of Post-War Building, Standards Committee, the Women's Group on Public Welfare; and there were a number of other committees to which the Society had given evidence (*SWHM Bulletin*, June 1943). It is clear that the Society's involvement in the work on reconstruction of housing was fairly extensive and time-consuming, and that it reflected extensive interest among women about the shape of housing after the war. What is less certain is the effect of this involvement. The response, for example of the Dudley Committee, does not seem any more understanding of the need for women's views to be taken into account than that of equivalent committees after the First World War (see Chapter 3 and McFarlane, 1984).

CONCLUSIONS

Women's employment in housing

It has been established that, up to 1939, women were employed as housing managers in a very limited number of organisations. The war, by creating a scarcity of male labour, provided the opportunity for women to be employed in a wider range of organisations in senior positions.

The nature of the change brought about by the war should be carefully considered. In the 1920s and 1930s women had steadily struggled through the ignorance and prejudice which might prevent them from employment in the newly expanding field of housing management – the wartime development can be seen as a continuation of this. Similarly, discrimination against women continued in some places during the war. However, it does seem likely from the evidence that the relatively sudden shortage of male labour made people who would not previously have employed women consider them.

This greater use of women's labour during the war happened in many occupations. It seems on the face of it to offer good support for the 'reserve army of labour' concept, 'the idea that women workers are particularly useful to capital as a reserve army of labour – to be brought in and thrown out of wage labour as the interests of capital dictate' (see Beecher, 1978; Adamson *et al.*, 1976). In his original formulation of this concept Marx was concerned to show how poorer and less skilled workers bore the brunt of unemployment, but the expansion of capitalism inevitably drew more and more people into a labour reserve of potential, marginal and transitory employment. Marx did not specifically consider women. Many later writers have felt that this concept is useful in interpreting changes in women's employment but it has also come under attack as being an undue simplification (Bruegel, 1979; Somerville, 1982).

As far as housing management is concerned, it seems that specific circumstances, particularly the Blitz and the management of MOS factory estates, highlighted the competence of some of the women who were employed and induced more decision-makers to support their employment. This corresponds with Somerville's (1982) argument that there are multiple forms of discriminatory practices and the reserve army of labour concept is too ambiguous and too broadly applied to be of much use. However, if there is predictive strength in this idea of women as a reserve pool of labour which can be disposed of at will, we should find women losing some of the gains made during the war when the war ended. Later chapters will continue to examine this idea.

Rubery (1988) has pointed out that there are three basic hypotheses which can be used for interpreting trends in women's employment. These are the flexible reserve or buffer hypothesis already referred to, the job segregation hypothesis and the substitution hypothesis (women being substituted for men as a cheaper form of labour). We have already had indications that the job segregation hypothesis is important for women's employment in housing; the substitution hypothesis could have backing from some of the events of the 1930s, but is modified by the fact that at times the women managers, because of their training and standards, could be more expensive than men. These theories will be discussed later in relation to the events of the 1960s and 1970s. But it seems likely that, as Rubery (1988) and Humphries (1988) argue, the theories are complementary rather than competing.

Women housing managers had made substantial gains during the war. They gained more employment in larger organisations and a wider range of work. A number of interviewees considered that their experience during this period played a crucial part in their careers – for example providing the basis for work at senior level in the New Towns. It was going to be more difficult in future for opponents to argue that women could not do this type of work.

The role of the Society

It would be possible to look at the war period as a time of considerable achievement by the Society and advance for women housing managers. The Society continued to play a vital role in supporting the employment of women managers. It succeeded in maintaining its organisation and membership despite the problems of wartime organisation.

This social support was further enhanced by the creation of the Junior Organisation, which was active over this period as a focus for the younger members. Meetings, committee work and the bulletin were also important in the dissemination of information to members. Despite all the difficulties of operating in the war period, by the end of this time the membership of the Society had actually increased substantially and the number of students had doubled.

The Society continued its direct work on employment, taking opportunities to publicise the work of women managers. There were still numerous authorities who regarded women as a temporary substitute for men, so the Society's work in protecting employment conditions for women was clearly vital at this stage. It had to concern itself not only with the level of salaries but also with the duties and responsibilities of the posts offered. But, despite the fact that women were still clearly often regarded as second best, the war provided an opportunity for a number of able women to take posts at central and local government level. This was of significance both individually, in giving those women a good chance of making further progress after the war, and collectively since it demonstrated that women could do the more senior jobs well.

On the other hand, it would not be appropriate to paint too rosy a picture about the gains made. The Society's numbers were still small. The restrictions of the training scheme meant that new members were added only slowly, and this was already causing some concern and shortage of qualified staff. The shortage of finance for training restricted members and the social class of new entrants. One cure for this which was advocated was the increase in trainee posts in local authorities, but this would effectively tend to take selection out of the hands of the Society. Moreover there were still only a few women in senior posts and some of the new appointees had been given temporary appointments or restricted to specialist areas such as requisitioning.

The influence of the Society was not quite so strong with the local authorities, which were still very male dominated, as it was at the Ministry of Health. For example, the Committee on Wartime Housing Problems, which was important as a channel of communication between the Ministry of Health and the local authorities, did not contain any women, let alone any Society members (Committee on Wartime Housing Problems, 1943b: 1). The history of women's involvement in the reconstruction planning makes it clear that it was as difficult for women's voluntary groups as it was for the women professionals to get their views properly taken into account and casts light on the continuing struggle to obtain gender balance in the formation of housing policy.

There was concern about what would happen after the war when everybody anticipated an increased demand for housing staff which the Society's training could not cope with. The debate about the admission of men reflects this and foreshadows one of the Society's major problems after the war. It might seem no longer appropriate to continue as a separate body but, if men were admitted, would the women be swamped and lose out in the contest for employment? Society members were aware that gains made in wartime would not necessarily be continued in peacetime. Signs of many of the problems which were to face the Society after the war were already apparent.

NOTES

1 H. U. Willink had been called to the bar in 1920. In June 1940 he was returned as Conservative and National Government member for North Croydon. Three months later he became Commissioner for the London Region, and held this post until October 1943. In November 1943, he was appointed Minister of Health and remained in this post until 1945, working on the preparatory work for the setting up of the National Health Service (Roberts, 1978).
2 Margaret (Peggy) Hurst had been the first housing manager in Cape Town in 1934 (see Chapter 5); subsequently as Mrs Margaret Hill she became first housing adviser at the Ministry of Health. References in this chapter and later are given in the name of Hill.
3 This was, at that stage, the Central Housing Advisory Committee's Sub-committee on the Design of Houses and Flats, a title later changed to the Design of Dwellings; often referred to in the literature as the Dudley Committee after the name of the chairman.

7 Post-war expansion

HOUSING POLICY, HOUSING ASSOCIATIONS AND HOUSING MANAGEMENT, 1945–65

Policy

The Labour government, which returned to power in 1945, inherited an acute housing crisis. Cullingworth estimated that the shortage of dwellings was in the region of 1,400,000 because the increase in population and in smaller households had added to the building shortage caused by the war (Cullingworth, 1960: 34).

Much thought had been given during the war to problems of solving the post-war housing crisis. The main thrust of the government's policy was on new building using generous subsidies to encourage the local authorities. Restrictions on private building, Town and Country planning and new high space standards were other important aspects of policy (Burnett, 1978: 281, 282). The success of this policy was jeopardised by financial factors and, in recurrent crises from 1946 onwards, the local authorities were encouraged to limit their housing output (Donnison, 1967: 166). The other major innovation of the Labour government was the New Towns programme. Fourteen New Towns were designated by 1950, mainly in the South-East. New Towns were managed by development corporations appointed by central government (Donnison, 1967: 306).

The Conservative government elected in 1951 continued to encourage local authorities to meet the expanded targets for house building, while at the same time beginning to relax controls on private builders and stimulate the private sector (Merrett, 1979: 246). But higher numbers were obtained by lowering standards (Burnett, 1978: 284, 285).

The period 1953 to 1956 saw a switch to supporting private

enterprise as the main vehicle for expansion, a restriction of local authority building for general needs and a return to large-scale slum clearance. Economic problems brought about further public expenditure and housing cuts in 1958 and 1960.

Housing associations

By the end of 1945 membership of the National Federation of Housing Societies stood at 210 housing societies and associations. The concentration on local authority building gave them little scope and the older housing associations found their economic position precarious until the Housing Repairs and Rents Act of 1954 (Allen, 1981: 51). However, the older societies did continue improvement work (interviewees). Self-build housing societies developed but their numerical contribution was small. The Coal Industry Housing Association was the largest of a new wave of industrial associations (Allen, 1981: 54) but another vigorous and prominent one was the British Airways Staff Housing Society, formed by employees themselves (Allen, 1981: 55). This became the unusual (for the UK) example of a large co-operative employing professional staff. In general, however, housing associations were managed along traditional charitable lines.

By 1961 the Conservative government, concerned about the shrinkage of the privately rented sector, gave attention to increasing the role of the 'third arm' of housing provision, initially by increased funding. In 1964, Central Government set up a new body, the Housing Corporation (Allen, 1981: 60–63). At first the Housing Corporation was concerned only with cost rent and co-ownership societies but it marked a new stage of government intervention in the housing association movement. Rapid expansion in this type of society followed, but their constitution and way of working was often very different from that of the older trusts and more akin to commercial bodies.

Housing management

It might be assumed that the crucial importance of housing in 1945 and the subsequent housing drive would have resulted in greater attention and resources being given to the local authority housing service. To some extent this was true, but the Ministry of Health continued to be responsible for housing, while it was managed by 1,700 local authorities. Mollie Empson, a SWHM member, succeeded

Peggy Hill as Housing Management Adviser, but one post was a very small commitment in terms of the Ministry as a whole (Fox, 1974). Any Central Government concern for housing management was expressed only in the monitoring of numbers built and in advice, either through the Housing Management Adviser or once again through committees and reports.

The Balfour Report (CHAC, 1945)

This will be considered briefly as an example of Central Government work on housing management and the changes in the role of SWHM. In January 1944, the Central Housing Advisory Committee set up a sub-committee to consider the management of municipal estates (CHAC, 1945: 3). They were asked to give particular attention to temporary accommodation. Unfortunately it is not possible to find much information about the workings of this committee since the relevant file (PRO HLG 37/20) at the Public Record Office appears to have been swept bare of everything except one letter. The introduction to the published report does mention specifically that the committee received oral evidence from the Society of Women Housing Managers and the Institute of Housing, as well as written evidence from a number of bodies (CHAC, 1945: 3).

The report, however, was not especially favourable to the views of SWHM. It is quite a brief document and began by noting that a number of points from the first report (CHAC, 1938) were still valid; for example, the concentration of the housing functions in one department was still very desirable and should be implemented (CHAC, 1945: 4). It was on the issue of the shortage of trained housing managers that the report was probably most disappointing to the Society. The final recommendation simply said: 'Local authorities should grant the widest possible facilities for the training of student housing managers under competent officers of their Housing Departments' (CHAC, 1945: 15) and suggested that arrangements for paying students should be extended. Once again no official help or money generally was made available for the training of staff. Moreover, in its text, the report had mentioned that 'professional bodies, such as the Chartered Surveyors' Institution, are interested in the requirements of the housing manager, and ready to meet them, if necessary by new developments' (CHAC, 1945: 11). By implication this might seem to devalue the Society's and the Institute's qualifications. The 1946 Annual Report of the Ministry of Health similarly mentioned the need for local authorities to create more training facilities and the fact that the CSI had decided to create a new qualification of high standard in housing management, and hoped that larger local authorities would be able to provide extended training facilities up to the standard of the new examinations (Ministry of Health, 1946: 168). Once again the Society and the Institute were ignored.

The pressures of housing management

If at Central Government level attention was on numbers and housing output, this was reflected at local government level. Housing management remained of low status relative to architects' and surveyors' departments. Glamour and prestige went to the departments producing the new building; the humdrum work of managing it still had a very low profile (information from interviewees; see also Dunleavy, 1981: 139). This had important repercussions on the 1950s and 1960s building programmes.

Interviewees who had been working for metropolitan departments testified that they were subject to the most appalling pressures in facing the pent-up demand for housing. Sometimes their task had not been made easier by the actions of Central Government:

'The dreadful thing was that, before we had in fact housed the last lot from the last rocket, the war was over and men began to come home and into the office saying "Where's my house? We've been told we're going to be housed when we come back from the army." They'd been shown prefabs in Delhi and Cairo and they'd been given a little buff form and they'd been told, "Fill up the form and take it back and you'll get a place like that." And there they were, in our office. More thumping of the table; worse than that. I mean, we had a chap jumping over the counter once, and a chap turning up (the housing manager's) desk onto her. And it really was terrible, because they simply didn't believe us, that we hadn't got houses for them.'

(Interviewee)

Such pressures on housing departments continued well into the 1950s. It does not seem surprising therefore that staff in housing departments developed defensive attitudes and strict 'rationing' theories in which they quite often followed popular prejudices (against non-residents, for example). Little academic attention was given to housing management during this period. A rare early example is the work of Kilbourn and Lantis which, as early as 1946, showed that a very wide gulf existed between the problems that housing officials considered the most important and those which were most significant to the tenants (Kilbourn and Lantis, 1946). Later critics of housing management outlined the problems they saw but seem to have given little attention to structural influences on housing managers' behaviour. This can be seen in the work of Burney (1967), Morris and Mogey (1965) and Sharp (1969).

Lambert, Paris and Blackaby (1978) similarly discuss the process of allocation in the context of housing shortage but say little about the effects this has on the attitudes of the staff. Cullingworth (1963) was one of the few writers who gave even brief attention to the working situation of housing staff. One of his case studies is discussed in more detail later in the chapter.

The central/local relationship in housing

One important administrative development of the 1950s was that housing was at last separated from the Health Ministry. The Ministry of Housing and Local Government became responsible for four main functions: public health, housing, planning and local government (Sharp, 1969: 15). The regional office organisation which had survived from the war was merged with the regional organisation which had been set up by the Ministry of Town and Country Planning.

Despite the fact that there now was a Ministry of Housing, its philosophy was still very much to leave management to the local authorities and not interfere – even in determining the amount of house building (Griffith, 1966: 220–295). Evelyn Sharp, writing in 1969, was prepared to criticise housing management as a whole:

> There is almost as much complaint about local authorities' general management practices as about their rent schemes. Readers of the popular press know how frequent are the stories of council tenants who are not allowed to keep pets or hang washing out to dry, and of tenants who are alleged to be in danger of imminent and unreasonable eviction or to have obtained a council house out of turn by favouritism. The number of such stories which attract the attention of the press is but a fraction of those which come to the notice of the Ministry.

> (Sharp, 1969: 87, 88).

But Sharp was prepared to fall back on the traditional line that this is not the Ministry's responsibility. 'Here, if anywhere, is a field where the local authorities are the absolute masters of their own policies' (ibid.), although she did admit that such independence was being breached by legislation such as the Race Relations Act 1968. Also it was clear that the Ministry had been active in offering advice and in investigating complaints:

> Complaints are so numerous that investigation takes up an unexpectedly large amount of the housing division's time. . . . The Ministry employs a full-time adviser on housing management – a qualified member of the Institute of Housing Managers – almost the whole of whose time is spent advising local authorities in the field and acquiring information for the Ministry about local practices, which are surprisingly varied.
>
> (Sharp, 1969: 88, 89)

There was no suggestion, however, that something more fundamental should have been done about such a volume of complaint. There was still only one housing management adviser, the housing profession's voice was weak at this level, and the voice of tenants was even weaker (Fox 1974; Dunleavy, 1981).

Growth in the size and scope of departments

By the 1950s, because of the increase in council house building, many departments had enlarged their size. In the 1960s inner London and metropolitan departments tended to be involved in slum clearance on a large scale. In addition local authorities, at the urging of the Ministry, had become heavily involved in system-built and high-rise building (Berry, 1974: 86; Dunleavy, 1981, 34–81). It was during this period that criticisms of housing departments began to be more generally substantiated.

It was alleged that housing departments often handled their responsibilities in an insensitive way but, sometimes, the critics recognised that housing departments were partially the victims of other departments' or government policies. Muchnick (1970), for example, studied the making and implementation of Liverpool's redevelopment policy in the 1960s. This was a case where the housing department did have enormous powers and was responsible for programming the clearance and redevelopment. There was a close interlocking of the interests of the housing department and those of the politicians, and lack of political representation of those actually moved by the redevelopment programme.

The growing scale and scope of housing work was an important influence. Slum clearance could be seen as a large-scale numerical operation and some authorities gave increasing attention to numerical targets and to ideas of efficiency which tended to emphasise quantity rather than quality (interviewee).

Another focus for criticisms was allocations policy. Elizabeth Burney's *Housing on trial* (Burney, 1967) was a critique of local authority allocations policy from the aspect of race relations. Burney pointed out how the factors which led to direct or indirect discrimination against coloured applicants stemmed from the nineteenth-century roots of public housing and especially the idea of 'grading' tenants, which people tended to associate with the 'Octavia Hill' tradition. Morris and Mogey, writing in 1965, were highly critical of housing management's claims to be anything more than a landlord and argued that housing management had neither the time nor the talents to deal adequately with social help, consumer research or community organisation as some housing management apologists claimed it did.

It was coming to be realised, as Morris and Mogey mentioned, that 'standards of house care have risen markedly since the war, and that the proportion of ''problem'' council tenants is now much smaller'. The steady growth in the professionalism, complexity and organisation of social work since the 1930s had made housing claims to deal with such tenants vulnerable. Morris and Mogey felt that councils had very stereotyped views of tenants and commented on the paternalism of housing departments and the ways in which tenants' freedom to enjoy their own home was limited.

Donnison (1960) ended his review of housing policy with a brief discussion of the uneven nature of housing administration in the local authorities. He expressed doubts as to whether they could respond even to a wider view of housing management, let alone become comprehensive housing authorities. His summary makes interesting reading.

> In attempting to introduce criteria of social justice and administrative efficiency into the field of housing management – a field previously governed by the entirely different criteria of personal influence, local tradition, profit and loss – the housing authorities are carrying out one of this country's most difficult and important experiments in social administration.
>
> (Donnison, 1960: 35)

Despite doing less than justice to some pre-war local authorities' and housing associations' efforts to house those in need, the extract does express the considerable misgivings with which many in the social administration field viewed housing management.

A case study of change

One example of local change in housing administration which did receive a brief public mention in the literature was that of Lancaster. J. B. Cullingworth in *Housing in transition* (1963), includes a short account of events in the housing department, beginning with the enormous pressure after the war:

> These were very stormy years for the local Council. Housing had become a political issue of enormous local importance. The weekly newspapers carried full reports on housing matters and pressed the Council strongly for 'full information' on the operation of its policy.
>
> (Cullingworth, 1963: 38–39)

The expansion of the housing programme to meet these needs and respond to government policy by 1953 resulted in concern about the costs of the housing programme, and the necessity for further rent increases led the Council to look carefully at its costs:

> The only item which seemed capable of being cut was that of management. The Housing Department, set up in 1934, had developed into a large welfare organisation with a staff of sixteen, seven of whom were qualified housing managers. The Department was run on Octavia Hill lines and was approved as a training office by the Society of Housing Managers.
>
> (Cullingworth, 1963: 48–49)

It was decided that a reduction of the annual cost of management from £4 to £2.10s a house was to be achieved by transferring responsibility for repairs to the Engineers' Department, dismembering the Housing Department and giving notice to all but two of the staff, bringing in new untrained staff instead. Cullingworth comments that 'These issues were confused by certain clashes of personality which make it difficult to give a clear analysis of the situation' (ibid.: 49), but it is at least clear that there was a sweeping away of intensive management. Although it is claimed by Cullingworth that costs fell abruptly, this was only from £11,000 in 1952–53 to £8,900 in 1954–55; this does not quite seem to be the order of saving promised. There is no indication of the effect on the quality of housing management and outcomes. There is an absence of other local examples as well documented as this.

GENERAL TRENDS IN WOMEN'S EMPLOYMENT

We have seen that the demand for labour during the Second World War meant that women's participation in paid work generally and in housing increased. This was regarded as a temporary measure and at the end of the war there was pressure for men and women to return to their traditional roles. Some writers have stressed the extent to which women's participation was rolled back and an attempt made to re-establish traditional values. There was a great stress on 'femininity', especially in the 1950s (Birmingham Feminist History Group, 1979; Riley, 1979).

The 'reserve army of labour' concept has also been applied to the disposability of women's labour after the war. However, an economic crisis in 1947 revived the call for women to return to work, though once again this was seen as a temporary measure. Wilson argues that the stereotype that contemporary feminists have developed of women being driven back into the home after the war is incorrect: 'Actually women were drawn into the labour force in growing numbers' (Wilson, 1980: 188). Joseph (1983) explored the trends in women's employment up to the 1970s. Working from official sources he demonstrated a long-term trend of increase in women's participation rates. He argued that female participation was rising over the whole of this period, apart from the under-24 groups where there was a slight decline. The other major trend was that the contribution of single women, who had been the major part of the workforce before the war, declined, being replaced by increased participation of married women in the workforce (Joseph, 1983: 68–77).

Women's employment in housing

It might have been expected that the attention given to public and social housing after the war and the expanding scope of activity would offer new opportunities to women housing managers. Interviewees confirmed that the outcomes were less certain than this and that they had to cope with a number of problems. First of all there was the issue of men returning from the war and expecting to be reinstated in jobs that women might have occupied. In a few cases this did result in women losing jobs but the wisdom of the Society's wartime policy of encouraging women to concentrate on what were basically new jobs reduced this risk.

More substantial problems arose from overt or covert discrimination. As housing departments became bigger and more prestigious there was more competition from men and the presupposition at the time was that men were better at larger-scale management. A heavily quantitative and cost-cutting emphasis could be used locally, as it was at Lancaster, to oust (female) trained management. Chapter 9 discusses this issue further and examines some limited statistical evidence. The lack of strong academic input into housing education meant that the academic critique of housing management was only slowly emerging and tended to be seen as divorced from practice. It had certainly not begun to pick up issues of gender inequality and it took nearly ten years for Burney's 1967 comments on housing allocations and race to be followed up by more substantial statistical studies. Tenants were beginning to become more vocal but not yet having much impact. So local authorities were free to do their best or worst with housing management as they thought fit.

Many of the younger generation of Society members were able to benefit from improved opportunities post-war by moving to those authorities which were prepared to employ women managers. Society's increased emphasis on 'femininity' and the tendency for women to want both a career and a family produced additional challenges. But it is clear from interviews that the women attracted to housing work were often trend-breakers of some determination. They might be impeded by society's prejudices but they were not deterred. Thus neither the 'reserve army of labour' theory nor the 'substitution' theory fit exactly in housing. The reality was more complex and locally varied.

THE SOCIETY AND THE ADMISSION OF MEN

The basic arguments

Chapter 6 has already described how the question of admitting men had been raised as early as 1943. The main argument put by Jean Thompson, who proposed the motion in November 1943, was that it was in the interests of housing to admit both men and women; if the Society did not, then men would have no alternative but to enter the Institute of Housing and 'if we insisted on a one-sex training we might hold up a big advance' (D.W., 1944: 4). The arguments put forward against the resolution gathered around the danger of swamping the Society by admitting large numbers of men, men getting the best posts, overloading the training system, and the loss of the all women's society (D.W., 1944: 5).

A letter from 'a mere male' in a subsequent issue of the *Quarterly Bulletin* put the challenge firmly in SWHM's court:

> What are these qualifications which make women exclusively fitted for housing management about which we hear so much? Will you be able to persuade the Ministry of Health to exclude men from the national post-war schemes for housing, whilst you are at the same time protesting against sex discrimination and claiming equality?
>
> (A Mere Male, 1944)

It was not surprising that the question was raised again almost as soon as war ended. As one of the Society members most involved in the admission of men said:

> 'It was so clearly absolute nonsense not to have them. I mean, there we were with the post-war boom in housing . . . and this tremendous need for proper housing management and yet we were, with what we said was a marvellous training scheme and all our wonderful tradition, we were limiting it to women – and in those days it was mostly unmarried women because in those days married women didn't – a lot – come forward for training. It was blatantly stupid.'
>
> (Interviewee, member of Unification Committee)

On the other hand, there were still a lot of misgivings in the Society about admitting men, especially among the older members:

> 'As far as I remember, it was mainly the younger people who wanted men to be admitted and the older people who were against it. And you couldn't blame them, because they'd had to fight, tooth and nail, for recognition and I think they probably foresaw the way we should be taken over by the men.'
>
> (Interviewee, member of Unification Committee)

> 'I can remember Miss Samuel pleading, with tears in her eyes "If you let men in, they'll take over all the best jobs, they'll be the Directors of Housing, you'll be the rent collectors".'
>
> (ibid.)

Younger members, especially those already involved in local authority careers, had less sympathy with this point of view.

> 'The Society missed the bus. I don't think Octavia Hill would have stood for a moment in keeping it for women only for as long. . . .

After the war, my generation were longing to open it up because there were hundreds of dedicated young men coming out of the forces who wanted to come in, and who had no respect for the Institute because, although they ran an exam, nobody was trained. . . . Anybody could get in, they were terribly lax over admitting.'

(Interviewee, member of Unification Committee)

As there was such a wide division of opinion the decision on the admission of men took some time and was, for the Society, quite bitterly fought. At the 1943 AGM the motion was first put:

That in the view of the immense development of housing schemes likely to take place after the war and the public recognition of the value of trained management, the Council be asked to consider the possibility of making the Society's training available to men as well as women.

(D.W., 1944: 4)

This resolution was passed by a large majority, only 10 votes being cast against it (D.W., 1944: 5). But, although the motion referred the matter to Council, there is no formal record of discussion until July 1945. By 1945 both members internally and the National Association of Local Government Officers (NALGO) externally were pressing for the Society to take action on this (SWHM minutes, 18.8.45).

At the AGM in November 1946, a motion instructing the Council to take appropriate steps to amend the Memorandum and Articles of Association of the Society so as to allow the admission of men was passed (SWHM minutes, 7.12.46).

The Society was now faced with the task of trying to implement this resolution. An Ad Hoc Committee was set up to consider procedural matters. It met every month, working its way carefully through the constitutional and other changes that would be needed on the admission of men. It recommended that regional meetings should be held so that members and students would have the opportunity to discuss the proposal before it came up at the AGM. An opportunity was also provided at the Annual Provincial Conference to discuss this issue.

It became clear that 75 per cent of the members would have to be in favour of the resolution before such an amendment could be made. A memorandum on the action taken by the Council was circulated to all members before the Annual Provincial Conference and it was agreed that the special resolution would finally be put to the AGM in

November 1947. But the special resolution did not receive a sufficient majority; 146 were in favour and 56 against (SWHM, 1948: 4).

The matter was not left to rest. Presumably because the vote had been so emphatically in favour of the resolution, it was thought worth trying again. Forty-five Fellows and Ordinary Members called for an Extraordinary General Meeting to reconsider the issue, a call which the Council had to comply with, and arrangements were set in hand for this (SWHM minutes, 10.1.48). The Extraordinary General Meeting was held on April 18th 1948. The special resolution was proposed by Miss J.M.Y. Upcott, one of the people who had worked with Octavia Hill, and the motion was carried by 152 votes to 44. Those campaigning for the admission of men had made a wise choice.

Although it may not have been fully apparent at the time, the admission of men did leave the Society with a problem about its role. Hitherto it had had a dual function – the promotion of good housing management following Octavia Hill methods, and furthering the cause of women in housing employment. Once men were admitted to the Society, it could no longer promote women's employment in quite the same way. But many of the changes in housing were beginning to bring even more into question the relevance of 'Octavia Hill methods'. It was clear that methods would have to be adapted and changed; but if they changed, how far were they any longer distinctively the Society's? And if both aspects of the role changed, how far was there justification for the survival of the Society as a separate body? The extent of these problems gradually became apparent during the next few years.

THE SOCIETY'S EMPLOYMENT ROLE, 1945–65

In the pre-war period and during the war, the Society had almost acted like a trade union in working to protect the employment interests of its members. In the post-war period, circumstances changed and the Society gradually lost this role, at a time when the employment of women managers seems to have been under some pressure. The process can be traced by looking first of all at the Society's general role in regard to salaries and the employment of women and then examining the record of how certain cases were dealt with.

General work on salaries and equal opportunities

In December 1944 the Society had approved new salary scales for housing managers, as it had done in the past (SWHM minutes,

10.12.44). There had always been a certain amount of negotiation about these scales but this time there was more serious trouble about getting them accepted in the local authorities. Local authorities employing SWHM managers had been informed of the new scales but

> The Town Clerk of Hemel Hempstead had suggested consultation with the Eastern Provincial Council for Local Authorities Administrative, Technical, Professional and Clerical Services. Mrs. Cully (Cheltenham), Miss Samuel (Bebington) and the Town Clerk of Rotherham had stated that their Councils would only negotiate about salaries with such bodies, not with individual professional societies. The Town Clerk of Rotherham had suggested getting in touch with the Association of Municipal Corporations
>
> (SWHM minutes, 3.3.45)

In other words, the growth of more formal unionism in local government meant that the functions of a professional society and of a union were becoming more clearly differentiated. The Society decided to stick to its salary scales but also to make a tentative approach to the National Association of Local Government Officers to discuss the matter (SWHM minutes, 3.3.45). It was agreed to make Rotherham a test case for negotiating through NALGO.

Negotiations with NALGO were protracted and the issue of the exclusion of men from the Society was raised by NALGO. Procedure for the application of grades in the Administrative, Professional, Technical Division of the national scales with NALGO backing was gradually agreed, though there was still a problem about the grading of trainees (SWHM minutes, 11.5.46). The Society thus lost some of its role with regard to salaries although it still took part in these general negotiations and in pressure for equal pay (SWHM minutes, 22.4.45) and equal opportunities.

Action on individual authorities and casework

In the pre-war period, the Society had often been consulted by individual employers on conditions of employment for women managers and had also acted where managers had felt circumstances to be unsatisfactory, though it had not always been successful in this.

In the post-war period, there are reports of a number of incidents where the Society was unable to take any effective action. The fact that those cases reported in the minutes included authorities as diverse as King's Lynn, Chesterfield, Romford, Newcastle under Lyme, Guildford and Tunbridge Wells may indicate general factors affecting women's posts, and reports of two of these incidents will be examined briefly.

King's Lynn

The housing manager had asked 'whether the Society would support her application for a higher grading for herself and her deputy'. It was agreed that the Society could not be of direct assistance (SHM minutes, 14.1.61).

Newcastle under Lyme

A letter was received from a SWHM member at Newcastle under Lyme stating that, as the Council were considering changing over to a rent collector and welfare officer organisation, all members of the Society on the staff had resigned. The minutes flatly record that it was agreed that no action should be taken (SWHM minutes, 12.4.47).

The significance of the Society's changed role

The cases reported in the minutes provide some indication that the changes in the Society's role in respect to employment were quite significant. This was a period in which, as we have seen, the employment of women housing managers was to some degree under attack. These cases indicate that the Society was no longer able to give such effective support to its members in local employment issues as it had pre-war.

RECRUITMENT AND TRAINING

Recruitment trends

Initially, the membership of the Society received a boost with a number of new students coming in. The Annual Report for 1945–46 reported:

> The increase in the number of women applying for training which was beginning to be apparent in the summer of 1945 has continued, and the influx of students, most of them newly released from war service, created a record at the beginning of 1946.
>
> (SWHM, 1946: 2)

However, this welcome increase had brought its own problems since some students were having to wait for months before a training office could be found for them (SWHM, 1946: 2). The increase in the number of students was not maintained in later years (see Tables 7.1 and 7.2). The Society continued with its examinations, revisions of the syllabus and careful attention to placement in training offices. In addition it began to run more short courses and study schools for students.

The continued existence of the Junior Organisation into the post-war period indicates one means of general support for students and younger members. The Junior Organisation published its own broadsheet for some time (copies for the period 1949 to 1951 are extant and show a lively range of interest). It also contributed to the debates about training, but was struggling to maintain the interest of its members and concerned about the slow rate of recruitment (see for example, Clutton, 1949)

Finance for training

The Balfour Report of 1946 had failed to make any provision for finance for training for housing management other than recommending that local authorities should increase their provision of trainee posts. While this did ease financial difficulties for some students, it also meant that recruitment was less under the control of the Society and it was the local authority which decided which examinations the student would sit. Also the Society's preferred pattern of training in two offices, one of which was a private one, did not necessarily fit with the common pattern of local authority traineeships which assumed that the students would stay with the original employer. So the practice of insisting on two offices was one which came under review in this period (SHM, Ad Hoc Committee on Recruitment, Training and Membership, 1959). It was also clear that the Society should give more attention to the question of payment of students during training (SWHM, February 1948) but this was difficult when the Society was insisting on training in a private office where the attitude to paying students might be less generous than in local authorities.

Relationships with the RICS

The RICS acted as the examining body for the Women Housing Manager's Certificate. But in 1945 they introduced a Professional

Table 7.1 SWHM/SHM membership figures published in annual reports 1938–58 (No figures published in 1946)

Year	Fellows	Ordinary Members	Licentiates	Associates	Students	Totals	Fellows & OMs non-practising
1938	26	159		63	60	308	
1939	25	178		70	69	342	
1940	24	173		72	87	356	
1941	24	197		69	73	363	
1942	31	193		92	72	388	
1943	35	214		71	88	408	
1944	34	242		71	90	437	
1945	35	256	(a)	79	122	492	
1946	—	—	—	—	—	—	
1947	32	268	11	126	(b)205	642	48
1948	36	276	19	105	161	597	54
1949	36	296	29	111	136	608	52
1950	38	305	31	106	124	604	62
1951	39	323	38	108	117	625	63
1952	39	326	44	104	111	624	72
1953	38	329	53	99	113	632	68
1954	37	334	105	61	105	642	72
1955	46	325	64	106	88	629	68
1956	47	319	65	102	99	632	65
1957	46	329	62	114	65	616	61
1958	46	318	66	124	59	613	61

Source: SHM minutes 29th Nov. 1958
Note: Revised Membership Regulations came into force Sept. 1947 and the financial year was changed in 1952.
(a) Licentiate Scheme first in operation
(b) Heavy post-war intake

Associateship in Urban Estate Management, 'designed to supply the need of a professional training and the qualification for the management of the many large new housing estates which are coming into being'. This did not seem to be regarded with any alarm by the Society, who hoped that a number of their students would take this professional qualification (SWHM, 1946: 3). However, the fact that the Women Housing Manager's certificate did not provide any exemptions, even from the early stages of the Surveyors' professional examination (SHM minutes, 15.3.58), seems to indicate a lack of status wihin RICS for the certificate. The majority of RICS members were men (see Chapter 10) so this meant that candidates with the Housing Manager's Certificate were facing increased competition from well-qualified men. However for many years there were still few qualified candidates for jobs in housing; any kinds of qualification in housing management were thin on the ground (interviewees).

The Institute of Housing

The Institute of Housing qualification was the other main qualification taken by housing staff. Society members regarded this as being at a lower level than their qualification and pointed out that there was no practical experience requirement. The Society was therefore concerned when they discovered that their own qualification was being given less status than the Institute's in the Local Government Examinations Board List of Approved Examinations for Promotion Purposes, as this had implications for salaries. Remedying this involved long and protracted negotiations with the Local Government Examinations Board, including making arrangements for an independent assessment of the scripts of the two bodies by the RICS (SHM minutes, 16.1.60).

Finally, the Examination Committee of the LGEB recognised the examinations as parallel (SHM minutes, 18.3.61) but this was referred back 'because of possible repercussions with other professional bodies'. The Society expressed its concern (SHM minutes, 13.5.61) but by this time negotiations for the new professional unified qualifications were well on their way so the question was becoming less relevant.

Problems of recruitment

The main problem that seems to have arisen was the exacting nature of the Society's requirement for practical training. Certainly many interviewees felt that there were more opportunities for women in the post-war period which adversely affected recruitment for housing. The Society made efforts to intensify the drive for recruitment by getting articles accepted in national journals and increasing propaganda to schools and colleges.

The report on recruitment in 1959 suggested that some factors affecting recruitment were beyond the Society's control: population changes, the increased marriage rate among women, the tendency to marry earlier (which also led to women candidates choosing professions which offered more scope for part-time work), the change in the nature of housing work (involving more administration, 'which perhaps had less appeal to women') and a change in 'patterns of mobility', with girls being less willing to work away from home. This report stressed the need for a unified professional qualification and possible amalgamation with the Institute (SHM Ad Hoc Committee on Recruitment, Training and Membership, 1959).

AMALGAMATION WITH THE INSTITUTE OF HOUSING

The formation of the Standing Joint Committee

As we have seen, after the initial increase in membership around 1945 to 1948, the post-war period had been for the Society one of declining membership and re-examination of its role. Once the decision had been taken to admit men, the rationale for having two separate organisations concerned with housing work was severely weakened. The review of the work and development of the Society carried out in 1948 seems to have concentrated on maintaining its separate existence (SWHM minutes, November 1948). But relationships with the Institute of Housing had been improving. The next important step was the formation of the Standing Joint Committee.

This was established for a trial period of a year. Co-operation began on matters of common interest such as conditions of work and income tax relief on subscriptions (SHM minutes, 12.5.56). Some difficulties arose in 1958, but at a meeting in March 1959 the Institute pressed the question of amalgamation. They said that their main concern was that the status of housing management was weakened by having two professional bodies (as negotiations with NALGO had illustrated), and suggested that amalgamation would be worth sacrifices on both sides.

The Society's representatives said that there would be no point in having a single body it if was not a good one:

> We made it clear that, although we fully recognised that the best members of the Institute were as good as, or better than, the average in the Society, we were bothered about their tail (i.e. members accepted without qualifications). . . . We expressed our fear that, as the smaller of the two bodies, we would be swamped by amalgamation, but the Institute felt that this was not a danger because so many of their members shared our views.

The Institute members were then asked to give their views on the Society. The main points made were that they thought the strict 'Octavia Hill system' with a trained manager responsible for 300–400 houses was extravagant and unsuitable for a modern office. They wondered whether the Society's training scheme produced better results than the Institute's, felt that Institute members got better value for money because they had more branch meetings, and considered the Society's admission standards for membership to be too severe and inflexible.

The Society's representatives said the Institute's representatives' comments on the Octavia Hill system were out of date and it was unusual for any of their offices to be run in this way. But they considered the rest of the points fair comment (SHM minutes, 2.3.59).

By 1959 a more general debate had begun in the Society with regard to its future and, as it is well recorded, it provides a good opportunity for examining the differing views held by Society members.

The debate on the future of the Society 1958–59

The issue was first raised by the South Eastern Group of SHM. This group felt that, because of the small numbers of the Society and the fact that new entrants were few, the Society was no longer holding its own in the housing world. They had considered amalgamation with the Institute of Housing but felt that the Society would be swamped and its basic principles abandoned, so they were in favour of closer links with the RICS and asked for the observations of Council on this (Shaw, 1958). The Council, in its reply to the Group, felt that all regional groups should be asked to discuss the matter and submit opinions and suggestions (SHM minutes, 17.5.58). So a process of consulting the regional groups was set in motion. The reports from the meetings of the regional groups were received by the Council by the autumn of 1958 and illustrate a range of opinion (SHM minutes, 27.9.58).

The Scottish branch

This branch agreed that there was a problem, were unanimous against amalgamation with the Institute of Housing, favoured stronger links with the RICS, but were not sure how this would be viewed by the RICS. They discussed at length the reasons for the failure to attract students and some of these reasons are worth quoting:

> There is an impression that the Society may be following too closely on lines based on the book of Octavia Hill, which are not wholly appropriate now in view of the changes that have taken place in the housing sphere . . . some members felt that too much stress is laid on rent collecting. . . . It is not questioned that rent collecting is essential and most important and that it is necessary experience for anyone hoping to reach higher posts in housing, but there must be a clear distinction between the work of a 'rent collector' and the work of a 'housing assistant'.
>
> (Robertson, 1958)

London branch

This report contained a lengthy account of opening speeches by Miss Christopher (Marylebone Borough Council) and Miss Cockburn (Barnes Borough Council). Miss Christopher was strongly in favour of amalgamation with the Institute in view of the Society's difficulties and in order to get a united voice for the profession with less duplication of effort. She felt that the voice of Society members would be heard within the new organisation. Miss Cockburn was more dubious about amalgamation with the Institute:

> they had no selection, no clear educational minimum, no guaranteed training, no minimum housing experience. . . . The Society is alive, democratic, there are enthusiastic young members, the numbers have remained steady since the war and its influence is out of all proportion to its size. Therefore, there would need to be big and certain advantages in favour of the Society's disappearance and Miss Cockburn cannot see them at present.'

The discussion which followed summarised many of the points raised in other meetings: the desirability of having one body, doubts about the Institute's low standards, fear of being swamped. The majority felt that the Society should be preserved as it stood, increase its publicity and maintain friendly relations with the Institute of Housing (Philipp, 1958).

Stevenage Development Corporation

Members employed there thought that something needed to be done about the general attitude:

> In our attitude to housing and to tenants we must stop being patronising. . . . Historical circumstance had made housing seem a charity and housing managers the ladies bountiful. Working in a New Town quickly disillusions one in this respect. We shall accept [the Institute's] invitations . . . to their London meetings and go to see for ourselves what the 'ogres' are like when they are all together.
>
> (Members employed at Stevenage Development Corporation, 1958)

It is not surprising that ordinary Society members, not having been part of Council's discussion of finance and membership, should have been somewhat shocked to receive an invitation to discuss the Society's future; nor that the predominating view should be the hope that things could continue as they were. The views of the members at Stevenage were particularly interesting as they represent not only members working in local government (held by interviewees to be more in favour of one professional body) but also, because of the timing and nature of recruitment for the New Towns, some of the younger members; these were the members who were least apprehensive in their attitude to the Institute of Housing.

Council met to consider the group reports again on 29th November 1958. By this time statistics of Society and Institute membership had been prepared which tended to confirm the gloomier opinions (see Table 7.2). Council concluded from the consultations that there was no widespread demand for amalgamation with the Institute of Housing but that closer contact should be fostered. Reasons for problems in recruitment were again discussed and it was agreed to set up an Ad Hoc Committee on Recruitment, Training and Membership (SHM Minutes, 29.11.58; SHM Ad Hoc Committee on Recruitment, Training and Membership, 1959). In September 1959, the secretary reported on what was needed for a recruitment drive, and concluded that the Society should press for a unified training scheme and examination for entry into the profession. It also examined the reasons outside the Society's control for decline in membership.

The Joint Examination Board

There were good reasons why progress towards unification should first be made in the examinations. First the Society was concerned that with falling numbers the RICS might decide to stop running the examinations. Secondly, both the Society and the Institute were embarrassed by the dispute with the Local Government Examination Board over the status of the two examinations and during this dispute the nonsense of having two parallel examinations was pointed out to them. Thirdly, some pressure from Central Government was added to the pressure from local government. The Minister of Housing and Local Government, the Rt. Hon. Henry Brooke, MP, speaking at the Society's national conference in 1960 and following up remarks made about training in 'Councils and their houses', expressed 'concern that consideration should be given to introducing a unified professional qualification' (SHM, 1960: 1).

Following on this, a meeting was set up at the RICS to discuss the possibility of a unified qualification for housing management (SHM minutes, 14.5.60). Among the possibilities put forward by the RICS was the setting up of a Joint Examining Board to administer a new examination for housing management, with representation from the RICS, the Institute and the Society. It was agreed that a meeting should be held with the Institute to explore this, and this was done at a meeting of the Joint Standing Committee. Despite some difficulties, agreement was reached in a reasonably short time, a report on the constitution and rules of the Board being presented to the Society on 18th March 1961 (SHM minutes, 18.3.61).

The Society had been forced to cede the administration of the examination to the Institute and the provisions for practical training were not nearly as strong as the Society would have liked (interviewees). The Board was inaugurated on 1st January 1962 (SHM, 1962: 3) and proved an important vehicle of contact with the members of the Institute (interviewees). By breaking the link with the RICS, the negotiations over the new qualification thus provided an important precedent. It made it more likely that, if an independent Society had to be abandoned, the idea of amalgamation with the Institute, which had previously been resisted by the majority of SHM groups, might be seen in a more favourable light.

Table 7.2 SWHM/SHM selection sub-committee figures 1948–58 (including candidates in Northern Ireland)
('A' = Accepted for training; 'R' = Rejected)

Year	Female Full A.	Female Full R.	Female Licen. A.	Female Licen. R.	Male Full A.	Male Full R.	Male Licen. A.	Male Licen. R.	Total interviews both classes Female	Total interviews both classes Male	Grand Total
1948	34	3	1	2	–	–	–	–	40	–	40
1949	35	–	11	3	1	–	–	–	49	1	50
1950	53	3	17	1	1	1	2	–	74	4	78
1951	36	1	10	–	1	–	–	–	47	1	48
1952	35	10	1	–	2	–	–	–	46	2	48
1953	23	2	6	1	1	–	–	–	32	1	33
1954	27	4	8	–	4	1	–	–	39	5	44
1955	33	–	9	–	–	–	–	–	42	–	42
1956	20	1	9	–	2	1	1	–	30	4	34
1957	11	–	9	2	1	1	–	–	22	2	24
1958	17	–	6	3	1	1	–	–	26	2	28

Source: SHM minutes, 29th Nov. 1958.

The Nottingham conference

While the negotiations about the new qualification were still going on, the review of the future of the Society had been expanded. Council decided that the Regional Conference at Nottingham University 22–24 April 1960 should be wholly 'domestic', that the report of the Ad Hoc Committee and recommendations should be circulated for discussion, and that there should be group discussions on these issues for much of the time at the conference (SHM minutes, 28.11.59). Subsequently a list of questions was drawn up for the groups to consider. This structure and the increased information available to members seem to have aided a clarification of views for, at the final session of the conference, two recommendations to Council were made:

> To ask the Council to consider approaching the Institute of Housing about the possibility of forming a new unified professional organisation of those engaged in housing management.

> To ask Council to consider acting on the Secretary's report on recruitment and the recommendations of this conference as a matter of urgency.
>
> (SHM minutes, 14.5.60)

These recommendations from the Nottingham Conference seem to represent a decided shift in opinion from the consultation with the regional groups only a relatively short while beforehand. What was the reason for this? It seems likely that the length of discussion and the better information distributed had made members more realistic about the difficulties of continuing as an independent body. Possibly members were also forming a more realistic idea about the difficulties in increased links with the RICS. Evidence from the interviews indicated that senior members of Council were becoming more convinced about the necessity for amalgamation and, as in any organisation, their discussion with others would have carried weight.

The final stages

Owing to the absence of minutes for this period, the same amount of detail is not available for the next stages in the negotiations, but once the Nottingham resolution had been accepted and confirmed by a resolution at the 1961 AGM, matters began to move ahead.

   In the year 1962–63, the Standing Joint Committee considered the practicalities of unification and produced a report to present to the Annual General Meeting of both bodies (SHM, 1963: 2). The 1963 AGM gave its approval by a very substantial majority to the report and authorised the Council to implement the recommendations as soon as possible. Working together at this stage was particularly important for both organisations because of the imminence of London government reorganisation (SHM, 1964: 3).

Agreement was reached by the AGMs of both bodies and the new Institute of Housing Managers came into being in April 1965.

CONCLUSIONS

The period 1945–65 was a crucial one for the Society. Post-war housing conditions and post-war social changes both produced stresses which demanded change. The rationale for an all-female Society was increasingly bound to be challenged, but when the Society began to admit men it became increasingly difficult to carry on justifying the existence of two separate bodies for housing. With the admission of men, the Society lost its specific role of encouraging the appointment of women, while much of the work it had previously done in the employment field was seen as no longer appropriate in post-war conditions.

The growing critique of the implementation of housing policy and of housing management did not produce the kind of fundamental review of practice and training and commitment of resources which happened in other major services. Local authorities could cut housing management expenditure or rearrange departments with very little idea of the effects of their actions. The split between the two professional organisations and the influence of the RICS added to the fragmentation of a policy area which did not have institutionalised academic back-up like education and social services. The underlying Conservative reservations about state intervention in housing made consistent development even more difficult.

Although the Society had benefited from the post-war enthusiasm for housing, the increase in recruitment was not sustained for long. Central Government initially paid little attention to housing management and vague phrases in official reports about more trained staff for housing were not backed up by money. This meant that the Society's training pattern put financial pressure on students who did not have independent financial resources. Local-authority-based training places began to appear but these could equally well be available for Institute of Housing training, which was less exacting and expensive. Increasingly the only option seemed to be amalgamation with the Institute of Housing. This left the Society in a slightly weak bargaining position, although the Institute of Housing was also keen on the merger and to some extent made the running in raising the issue. From the Institute's point of view, the existence of two separate organisations was an embarrassment, especially in negotiations with national bodies and with Central Government and because of the Society's good reputation.

The move toward unification in 1965 seems to have been rather more trouble-free than the campaign for the admission of men in 1948. But it was not entirely smooth going, because in both organisations there were those who opposed unification. The Standing Joint Committee was in a position, typical of negotiators (see Rubin and Brown, 1975: 13,14), of 'fighting two battles . . . you were negotiating with the enemy so to speak and fighting a rearguard action with your own troops' (Institute interviewee). Both sides were anxious that the bulk of their membership should be carried along with them and that a further split should not result. The Society, whose members had the most misgivings, carried out a very full and lengthy consultation process which seems to have achieved the result it wanted and was typical of its way of working. There does not seem to have been any specific check on drop-out of members after unification and the statistics given in the next chapter do not clarify this point completely. Society interviewees thought that one or two of the older members may not have transferred to the new body, but they would have been about retirement age anyway. Certainly there was no wholesale defection from either body.

This does not mean that there were no doubts or reservations. Some of them have been stated or implied in the extracts already given; but in the next chapter we look at the advantages and disadvantages of unification as reflected in written and oral statements of those involved at the time, as well as in the statistics about membership after amalgamation.

8 Women in the Institute of Housing, 1965 to the 1980s

CHANGES IN THE COUNCIL OF THE INSTITUTE

The amalgamation of the Institute and Society had been achieved, but in the following years it seemed that the women had lost more than they had gained. Nowhere was this change more marked than in the Council of the Institute. It had been a condition of unification that representation of former bodies should be continued for the first three years of the new organisation, but after this women's participation dropped dramatically, as Figure 8.1 shows. The lowest point was reached in the years 1972–73 and 1973–74, when only one woman remained on Council. Representation of women began to climb again after that but still remained below the 1965 level. How far did this signify a reduction in women's membership of the new body?

CHANGES IN THE MEMBERSHIP OF THE INSTITUTE

Tables 8.1 and 8.2 summarise the changes for 1965, 1977 and 1983. They demonstrate that, even for members, the drop by 1977 was not as marked as it was on the Council and by 1983 some signs of revival were evident. The proportion of students rose steadily over the period. It looks as though the changes going on were complex. Because of variation in the trends between different classes of membership, it is desirable to look at these in more detail.

Fellows

Table 8.3 shows that both the total number and percentage of women Fellows declined markedly over the period. By 1983 the actual number of women Fellows was almost half what it had been in 1965 while the number of men Fellows had increased slightly.

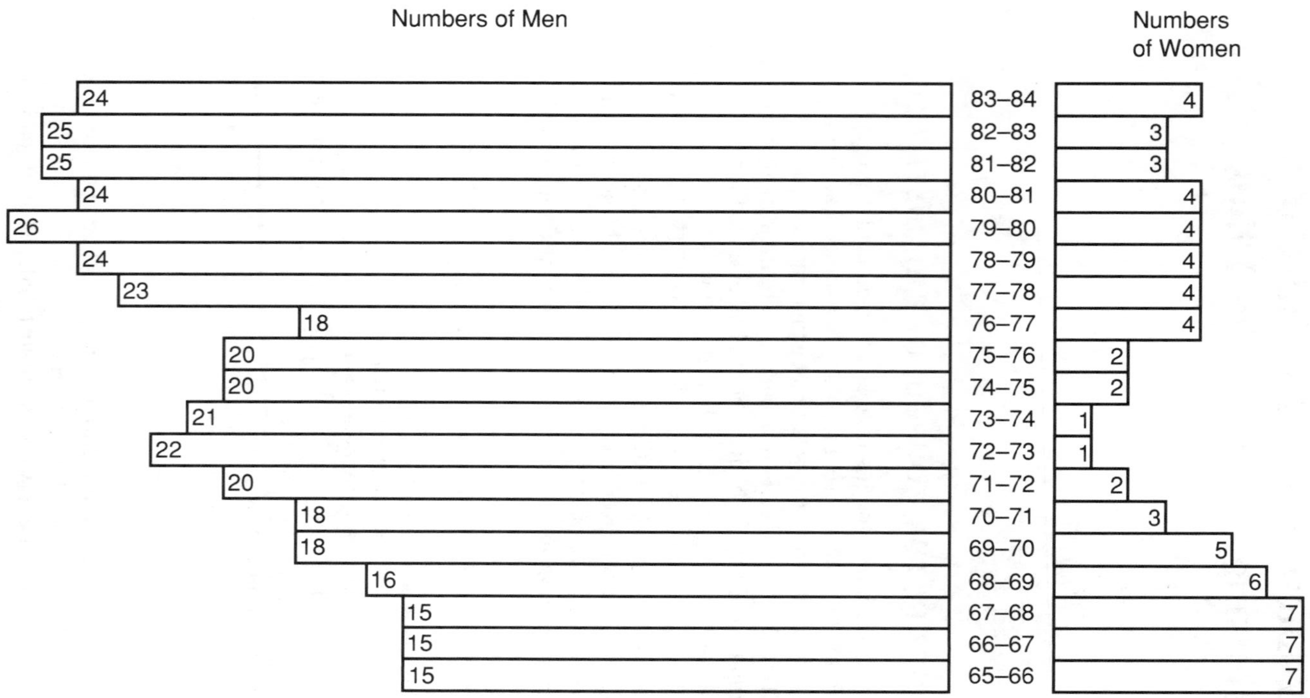

Figure 8.1 Institute of Housing Council – men and women members (1965–84)
Source: Institute of Housing Year Books and journals

Table 8.1 Institute of Housing members 1965, 1977 and 1983

	1965				1977				1983			
	Men		Women		Men		Women		Men		Women	
	N	%	N	%	N	%	N	%	N	%	N	%
Fellows	223	79	61	21	224	85	39	15	270	89	32	11
Members	713	79	189	21	952	82	201	17	1,250	77	367	23
Licentiate Associates	17	20	68	80	138	86	22	14	396	81	92	19
Total	953	74	318	26	1,314	83	262	17	1,916	80	491	20

Source: Institute of Housing Year Books and printout

Table 8.2 Institute of Housing students 1965, 1977 and 1983

	1965				1977				1983			
	Men		Women		Men		Women		Men		Women	
	N	%	N	%	N	%	N	%	N	%	N	%
Students	837	85	142	15	1,728	76	535	24	1,598	60	1,076	40

Source: Institute of Housing Year Books and printout

Table 8.3 Institute of Housing Fellows 1965, 1977 and 1983 by type of employer

	1965				1977				1983			
	Men		Women		Men		Women		Men		Women	
	N	%	N	%	N	%	N	%	N	%	N	%
All local authorities	200	87	30	13	168	87	26	13	212	94	14	6
Housing trusts	7		15		23		8		35		13	
Other public bodies	16		16		33		5		23		5	
Total	223	79	61	21	224	85	39	15	270	89	32	11

Source: Institute of Housing Year Books

Although the numbers of Fellows were small, changes here were particularly important because of the significance of senior women acting as 'role models' for younger women and in providing employment and promotion opportunities. This was also the group from which most Council members would be drawn.

The shifts in employment categories demonstrated in Table 8.3 are also informative. For local authorities the number of women Fellows in 1983 was half of what it had been in 1965. For housing

Table 8.4 Institute of Housing qualified members 1965, 1977 and 1983 by type of employer

	1965				1977				1983			
	Men		Women		Men		Women		Men		Women	
	N	%	N	%	N	%	N	%	N	%	N	%
All local authorities	637	86	102	14	775	86	128	14	962	81	221	19
Housing trusts	7	14	43	86	78	69	35	31	166	65	88	35
Other public bodies	69	61	44	39	99	72	38	28	122	68	58	31
Total	713	79	189	21	952	83	201	17	1,250	77	367	23

Source: Institute of Housing Year Books and printout

associations the number of Fellows only reduced slightly but the proportions were reversed, owing to a substantial increase in the number of Fellows employed by housing associations.

Members

The proportion of women members reduced by 1977 but the numbers remained stable (Table 8.4). From 1977 to 1983 the proportion of women began to rise again. However, housing associations (where women had predominated heavily in 1965) once again showed a reversal in 1977. Even in 1983 the proportion of members in housing associations who were women was only 35 per cent as compared with 86 per cent in 1965.

Licentiates and associates

These were the staff without examination qualifications and, because of changes in the definitions of this class, it is not useful to analyse the figures in detail. While 80 per cent of this class of membership were women in 1965 this fell to 14 per cent in 1977 and only improved to 19 per cent in 1983.

Students

In the period being studied only a small proportion of Institute of Housing registered students were full-time students at a polytechnic or university. For example, there were only two full-time degree courses for housing, numbers attending these were quite small and they were not included on the Institute pass lists.

The students of the Institute were, in the main, people in full-time employment. The majority were studying on day release at technical college with a minority studying by correspondence or on their own. The minimum entry requirement meant that their age would be at least 18 but the average age would tend to be higher, with some being in their thirties or forties.

Over this period, the Institute qualification tended to have a high failure rate. But once qualified, the student already had a job and practical experience and would be looking for career progression. Student membership and pass rate therefore has a particular interest for this study, because it formed the pool from which the qualified membership was drawn. If the proportion of qualifying students who were women was low, it would not be logical to expect a high proportion of women in membership. Until the 1980s the Institute

Table 8.5 Institute of Housing men and women qualifying 1965–83

Year	Men	Women	Total	Women as % of total
1965	25	12	37	32
1966	14	5	19	26
1967	25	8	33	24
1968	34	8	42	19
1969	83	14	97	14
1970	83	14	97	14
1971	144	22	166	13
1972	18	4	22	18
1973	69	22	91	24
1974	103	33	136	24
1975	112	31	143	22
1976	135	37	172	22
1977	167	83	250	33
1978	182	78	260	30
1979	199	101	300	34
1980[1]		incomplete data		
1981[2]	197	97	294	33
1982	145	91	236	39
1983[3]	32	61	93	66

Sources: Institute of Housing pass lists and *Housing Journal*
[1] Diploma pass lists for May and December missing.
[2] May Professional Qualifications only; diploma missing.
[3] First final pass lists to include the professional qualification.

did not monitor these statistics by gender, but it was possible to do so retrospectively with reasonable accuracy except for 1983 (see Chapter 11).

Table 8.5 shows very clearly a drop after 1965 in the percentage of qualifying students who were women, reaching its lowest point in 1971 with 13 per cent. After 1971 the rise is quite marked, up to 39 per cent in 1982. It should also be noted that, since *numbers* qualifying rose markedly over the period as well, a substantially larger *number* of women had qualified by 1983.

A possible reason for the increase in the percentage of women students is the introduction of the new Professional Qualification in 1981. Figures confirm the impression that the Professional Qualification did have a higher proportion of women students, though the percentage fluctuates.

It is worth noting that the figures for student membership of the Institute are less reliable indicators of the make-up of the pool of trained staff than the pass list. An inheritance from the old Institute of Housing was a large proportion of 'students' who were not actively studying. Comparison of the figures for students registered with the pass list information confirms that more of these inactive students were men.

The main trends

It can be concluded that there are two main trends to examine. The first is an overall long-term decline in women's influence and participation within the Institute from the mid-1960s to the mid-1970s, with some revival since. The second is a difference in trends for various groups, with more senior members (Fellows) and housing association employees showing the most marked and sustained decline and students showing the most marked revival.

It is worth noting that the amount of perceived change may have been greater than some of the actual changes. A comment by an interviewee was: 'The Society sank without trace.' Similar comments were made by other interviewees – who felt that women's influence was wiped out – and by younger members of the Institute, who saw it as a very male-dominated body at the time. This probably indicates the importance of the changes at Council and at Fellow level, since these were the people who were likely to be more prominent in Institute business and at public meetings. The ordinary members and students were not likely to be aware of the statistics of membership and in fact the breakdown by gender was not available.

The following sections of this chapter look at the reasons for decline in women's participation, particularly in the years following amalgamation. The focus is within the Institute, and especially on the changes at Council level, which we have already seen are of significance. First of all, the opinions of the interviewees on the reasons for the changes they perceived are examined. Then the literature on group interaction is examined to see what light this can shed on events. Finally, some of the literature on sex stereotyping is examined to identify the effect of these factors. The conclusion summarises the likely weight of these different factors.

But the changes within the Institute were also affected by changes in the world outside, particularly in housing employment. Because these changes are complex, they are examined in the next chapters.

REASONS FOR DECLINE IN WOMEN'S INFLUENCE AS SEEN BY INTERVIEWEES

In this section the main source of information is women who were members of the Committee on Unification. Comments from other interviewees and the male committee members interviewed are added where appropriate.

Difference in attitudes, particularly with regard to status

The most common reason given for the decline in women's influence, by five out of the eight interviewees, was the difference in aims and attitudes between the ex-Society and ex-Institute members, particularly with regard to status and style of meetings.

'I don't like generalising . . . but on the whole men tend to be more ambitious and they rather like being able to say, "I'm a member of Council". It always struck me that, when we got applications for jobs, the men always said . . . members of branch committee . . . etc., whereas women didn't think of it, really – didn't think of it so much as an honour or as an added qualification, more as something you bore.'

Eagerness to participate in meetings and the style of meetings was also felt to differ:

'If you take away the "club" atmosphere which was very enjoyable . . . I don't think that most of us are frightfully keen on lots of meetings and extra things – and of course they were a bit more formal, they do rather indulge in mayors and formal lunches and speeches and things, which isn't our style at all – much more status, wearing chains and things – I think men are like that, they like that sort of thing – look at Masons or Oddfellows . . . I don't think women do to anything like the same extent – and found it difficult to play up to that sort of thing.'

Another interviewee expressed the dislike for formality even more forcefully:

'The agreement we thought we'd found round the table was more apparent than real. Inasmuch as a lot of the things have crept back. A silly "for instance" – the Institute was full of the most stupid dangling of badges and chains of office and all this kind of thing and we cut all that out except for one specially designed badge for the President. And now it's got right back, every branch has it – and they got around it by making presents that people couldn't turn down. I mean we weren't going to have any "danglers" and then Mr X you see presented the President's badge.'

One interviewee felt that men were more competitive about getting on the Council:

'One of the most hotly contended points in the negotiating was this business of not sitting on the Council for ever. Now there was a different attitude. I mean in the Society when you were young you longed to be up there where the decisions were taken. When you'd been on Council say three times, and then it was Friday nights, Saturdays all day and Sundays up till lunch time . . . and a lot of work at home, you were thankful when your time came and you couldn't be elected, you had to have a wait, you had your year of freedom. In the Institute, they struggled to get on and once on they never thought of coming off and they just sat there for years.'

As the number of ex-Society members on Council reduced, the problem of a difference in attitudes became more acute:

'I was the last SHM left on Council, the last year. I was the only person of the old Society, I was the only person from a housing association and I was totally miserable. . . . It was impossible, really.'

The mechanics of election

A number of the interviewees had realised that the mechanics of election would work against any minority. But this was accentuated

by the constitution the Institute adopted, which gave heavy representation to the branches:

'You see you really had to do it through the branches and I think a
lot of our people weren't assiduous enough at the branches but in
any case . . . there were not enough of them – you see if you
imagine the Institute of Housing was four times as big as the
Society and you've got a branch which has never had anyone
from SHM in it before and by reasons of geography has only got
three members of the branch, or six or ten . . . and nobody who
votes has known them as long as they've known John or Reg – it's
as simple as that.'

Participation of women at branch level was sometimes problematic:

'And I think when we began to come to the surface again and
attended some of the meetings, the lectures and the meetings, talks
and discussions were of such poor quality, quite honestly, that I
personally felt it was a waste of time and I could spend my very
small amount of spare time better in other directions.'

One member felt that the image of the Society did not help women to
get elected in the new Institute:

'I think also that the reputation we had among Institute people was
of a lot of old fuddy duddies who didn't know anything about
anything and therefore weren't worth voting for. One got an
awful lot of jibes about Octavia Hill – I mean, a lot of it was
nonsense for those of us who were up and coming and managing
big departments and so forth. But one got the snipes back about
ladies with ink bottles strapped to their waist and so forth, it's
like the coals in the bath syndrome with tenants – a lot of that
came from the Institute, probably out of ignorance. But I think
that probably the two main reasons – that we just weren't known
or if we were known we were known for what we were, which
was little tight managing bodies which couldn't grasp the broader
issues – I don't think we'd ever pushed ourselves enough to try
to do it.'

Another ex-Council member felt that employment experience was a
factor: 'if the ex-Society people are not in the top jobs they are not
known'.

This tendency of the Institute to elect people with the 'big jobs'
was contrasted by some interviewees with the tradition of the
Society in encouraging participation of younger members. 'The

Society had more junior people on Council.' But one of the Institute negotiating team said: 'We don't want a lot of young girls on Council.'

It seems that a number of women, although they transferred to the new body and remained in it, did not actively participate, either because of unhappiness with the 'style' of the new body mentioned earlier or because of overload at work. This latter factor particularly affected some London members:

> 'Amalgamation was taking place in the London boroughs and those of us who were involved in that just went under and stayed under for about three years – just could not look beyond our own job responsibilities. And I think that is why quite a lot of people dropped out of active membership.'

> 'And I think rather than stay and battle with them on their own terms, we shrugged and turned away. That's how I feel and I know several others [names] battled on and they were marvellous. I can't speak for anyone but myself and my immediate friends, but we all felt we couldn't be bothered with it.'

The feeling that some of the women 'couldn't be bothered' or 'it was hopeless anyway' was clearly important in terms of women's participation. It was implied by some interviewees that there were a small number of ex-Society members who did not transfer or dropped out very quickly. Surprisingly enough, there is no extant report or record of this issue. But it is likely that the numbers of those who dropped out completely were small. Many interviewees considered that more women remained members of the Institute as a formal thing but stopped feeling that they 'belonged' or participating actively. This affected some of the younger members as well as older ones (e.g. D. Clark, personal communication). This is important because it explains why the proportion of women members did not fall as drastically as their influence on the Council did.

Intention?

Was there a deliberate intention to 'eliminate' the Society presence?

> 'I was discussing something with a Scots member who said "Ha, ha, you did not think about it when you amalgamated, that we would be two to one and we'd get everybody on." Well, I mean, we had thought about it and that wasn't our idea at all, we thought we'd vote for people on their merits . . . but if you're voting for

people on Council or something you do tend to vote for the people you know.'

(Interviewee, member of unification committee)

This suggestion that there was some intention on the part of ex-Institute members to deliberately crowd out the ex-Society members was only specifically mentioned by one other interviewee (not a committee member):

'I suspect it might have been deliberate policy. . . . I've no real reason to say that but I think in a lot of people's minds in the Institute this was a takeover and this was a deliberate policy to get the numbers down.'

This view was not shared by the ex-Institute interviewees from the unification committee. One admitted that he was 'not satisfied with the arrangements for democratic election. Once the safeguards had run out, two ladies who were admirable and carried responsible jobs were not elected. I wished we could have retained the brighter ladies of the Society.'

The reasons he gave were that there was a general feeling that they could not reserve seats for members of the old organisations and that old Institute members were always inclined to vote for people with the biggest jobs. So that the people who got on Council were 'men, and the people at the top of the profession. Inevitably, this did not take cognisance of the fact that these admirable ladies were not elected. I and some of my colleagues regretted that; they dropped out.'

Another ex-Society committee interviewee, while not going so far as to ascribe deliberate planning, felt that the wiping out of Society representatives was inevitable and she had in fact foreseen this. 'I was cynical; I'd been working with the men for so long, I knew what the London housing managers were like.' So she felt that whatever was said in the negotiations the majority would revert to their old ways once the unification was accomplished. She gave two examples of this:

'You knew jolly well that they were going to continue the system of having Council members going on for 20, 30 or 40 years; no matter what you said about having to come out every three years they'd find a way round it . . . You knew darned well that they'd go back to the Institute of Housing just as soon as they'd get the chance. I remember the arguments over the name.'

(The Society had insisted on the inclusion of 'housing managers' in the title but the name was changed back to Institute of Housing in 1975.) This interviewee summed up that the women lost out because 'they were all too nice, the men were all too nasty'.

Two other interviewees felt that Society members did not get an accurate view of what the Institute was like:

'I think we thought we were getting more say in it than we actually got. I think we underestimated how much the rest of it [the Institute] was. We looked only at the gilt tip and we didn't look at the tail very much and thought perhaps that tip was representative, but I don't think it was. We saw the best of the Institute and we didn't see all this other mess that was going on underneath.'

(Member of unification committee)

One of the ex-Institute interviewees agreed with this. He felt that inevitably the negotiating team might not have been totally representative of both organisations, but this was more marked in the case of the Institute, because in the Society

'Everybody was properly qualified and the educational background was fairly rigidly applied, I think, so that they were a more homogeneous body than we were so in a way it could be true. I mentioned earlier that the Institute had these widespread standards . . . a more varied body . . . a high proportion of the members were, I suppose, to put it bluntly, people who hadn't qualified, in some other department – treasurers or town clerks or something – and perhaps had been put on to some aspect of housing work, or sometimes the engineering department.'

However, most of those women who felt that the Society members had, in the end, lost out badly or that they had failed to recognise the true nature of the Institute membership, thought that the decision to amalgamate was still justifiable.

'We've been submerged, but would we have died anyway?'

'My two chief assistants are my support and . . . were all against this amalgamation, and a few other people were too; they said we shall just be buried. They were really right, but there was nothing else to do, we hadn't the money, we hadn't the resources to stand on our own.'

(Member of unification committee)

REASONS FOR DECLINE IN WOMEN'S MEMBERSHIP: GROUP DYNAMICS

The interviewees identified a number of potential problems in the process of unification – the differences in attitudes between Society and Institute members, the mechanics of election, the feelings of the women that they were in a minority and not accepted. Unification involved the mixing of two previously separate and somewhat rival groups and it is relevant to look briefly at the literature on relationships between and within groups to see what light it can shed on the processes involved.

Relationships between groups

The situation of friction between two groups is one which has been explored in the literature of social psychology. Earlier writers linked hostility to other groups (out-groups) with early childhood development or with the development of certain types of personality, e.g. theories about the 'authoritarian personality' (Allport, 1958; Adorno, Frenkel-Brunswick, Levinson and Sanford, 1950). Subsequently these were strongly criticised (for example, McKinney, 1973). Theories such as that of Sherif (1967) argue that existing relations between groups generate attitudes which are consistent with that relationship. Looking at the stages of group formation Sherif hypothesised:

> When a number of individuals without previously established relationships interact in conditions that embody goals with common appeal value to the individuals and that require interdependent activities for their attainment, a definite group structure consisting of differentiated status positions and roles will be produced.
>
> (Sherif, 1967: 76)

In this situation 'norms regulating their behaviour in relations with one another and in activities commonly engaged in will become standardised, concomitant with the rise of group structure' (ibid.).

A number of the earlier chapters of the present book describe the stages of the establishment of the Society of Housing Managers as a group with common goals, and the emergence of a group structure. The descriptions of the Society illustrate the strength of group norms within the Society, which was particularly close knit. I have not described the formation of the Institute in equivalent detail, but it is clear from interviewees that it had also produced its own structure and norms.

Sherif further hypothesised that

> When members of two groups come into contact with one another in a series of activities that embody goals which each urgently desires, but which can be attained by one group only at the expense of the other, competitive activity toward the goal changes, over time, into hostility between the groups and their members. . . . In the course of such competitive interaction . . . unfavourable attitudes and images (stereotypes) of the out-group come into use and are standardised, placing the out-group at a definite social distance from the in-group.
>
> (Sherif, 1967: 81)

These hypotheses have a reasonably good fit with events relating to the Society and the Institute. In the 1930s the groups were, as shown in Chapter 5, competing for the same goals: the employment of their members in senior posts in housing organisations, recognition by local authorities and by Central Government as the premier voice of authority on housing management, acceptance of their particular views on housing management. After the initial breaking of common membership of the Society and Institute considerable hostility developed between the two bodies. The rivalry between them continued during and after the war and it was only gradually, as the disadvantages of having two bodies became more pressing and more common goals appeared, that hostility diminished. This also is consistent with Sherif's theories, which argue that intergroup hostility will diminish only in the presence of superordinate goals requiring continued co-operation. But as Sherif (1967: 93) argues, the human past has a heavy hand. Intergroup attitudes must be interpreted within the framework of people's past relationships and their future goals.

Stereotyped attitudes arising from the past relationships remained to cloud relationships within the new Institute. 'Ladies with ink bottles strapped to their waist' is an example of an Institute stereotype of Society members (as related by the latter). One Institute interviewee said that there were 'people on both sides who took the most ridiculous attitudes for years . . . men, who felt that the Society members were "a lot of sloppy women who get all sentimental over housing"' (unification committee member). Another commented that when Institute members began to meet Society members more they found that they were not as 'prunes and prisms' as they had imagined. His comment that 'some chaps were sternly anti-feminist, you know' indicates the extent to which the Society was seen as feminist even though some members – and some feminists today – might reject this label. Meeting Society members, Institute members found they were not 'ladies, you know, of the kind which they envisaged' and 'it was an eye-opener' to realise the quality of some of the Society members. This Institute interviewee had some awareness of the psychological processes at work:

> 'There was a good deal of prejudice against a lady who was in charge at [X] at one time who was said to be rather aggressive . . . in these situations you get a lot of rationalisations . . . reasons which aren't reasons . . . what it really is is our party wouldn't be in power if . . .'

The Society stereotype of the Institute was less vivid but included the view that Institute members were less educated, did not care so much about people, that they were much more concerned with status and large numbers and did not have such high standards or professional practice. One Institute member described the Society stereotype of Institute members as he saw it: 'people had to meet across the table . . . they found we didn't bring a pint of beer with us.' Each group tended to underestimate the achievements of members of the other group; little recognition was given in interviews to pioneering achievements of members of the 'opposing' group.

The superordinate goal that had emerged in the 1950s and 1960s to bring these two conflicting bodies together was that of achieving a unified professional body for housing. It could be argued that it was by then clear that neither body could achieve its aims without the co-operation of the other. But not all members of either body were convinced that this goal was really worth the price. So the possibility of friction was high.

Sherif's ideas about the way in which incompatible goals are an important contribution to group conflict still stands (Gaskell and Sealy, 1976: 60–77). There has been confirmation that intergroup contact does not, by itself, necessarily reduce conflict (Deaux and Wrightsman, 1988). One factor in intergroup conflict which has been emphasised is the breakdown of communication between conflicting groups (Newcomb, 1947). This is also exemplified in the history of the Society and the Institute in the 1930s, and the post-war years show a gradual resumption of communications with the gradual perception of common goals.

Jamous and Lemaine (1962) studied competition between groups with unequal resources, noting that the handicapped group became discouraged and did not want to compete. This can be compared to the situation within the Institute after 1965, the Society being the group with fewer resources in terms of membership and the one which felt that, 'rather than stay and battle with them on their own terms we shrugged and turned away' (Society interviewee).

Relationships within groups

The literature on relationships within groups is enormous and only some relevant points can be mentioned here.

> Groups develop definite 'pecking orders' in terms of amount of speech and influence permitted. During the early meetings of the group there is a struggle for status amongst those individuals strong in dominance motivation.
>
> (Argyle, 1967: 70)

Low-status members of a group talk little. A person's position in hierarchy is primarily a function of how useful he or she has been in the past. The group uses techniques of reward and punishment to maintain this system.

> Groups develop norms of behaviour which can be regarded as a kind of culture in miniature. Such norms will govern the styles of social behaviour which are approved and admired. Anyone who fails to conform is placed under pressure to do so, and if he does not is rejected.

Thus Argyle summarises some of the facets of group interaction that have been studied (Argyle, 1967: 71). If we look first at the situation within the Council of the Institute, even at the beginning, with the guarantee for ex-Society membership, the ex-Institute group was numerically dominant. Ex-Society members were emphatic that the style of the new body was very much that of the Institute, partly because the secretary of the Institute, Henry Key, took over as secretary of the new body. Thus the ex-Society members were in the weaker position to start with. Whenever an issue came up which activated conflict between the norms of the two pre-existing groups, the Institute norms were likely to prevail, giving the ex-Society members even more of a sense of being outnumbered. As time went on and the ex-Society membership was reduced, these members, if they disagreed with the majority, were increasingly likely to be treated as deviant and rejected.

> I was the only person of the old Society, I was the only person from a housing association and I was totally miserable . . . it was impossible really

is a very accurate description of somebody regarded as deviant by the main group on at least two counts.

Ex-Society members in Institute branches would have been in a similar position. Their background and values were different and they were newcomers to the group. In each branch there might be only one or two women. Given the difference in norms, the likelihood of their feeling accepted was low. There was a difference in the ability or willingness of individuals to cope with this situation. Those members, particularly younger ones, who already worked in local authorities with men and co-operated with them in other housing activities (e.g. working parties), saw their future in those terms and would be, as they put it, more 'hardened' to this kind of interaction and to fighting their way in this kind of group. Others, who were perhaps nearer retirement, worked for housing associations and for a long time in a predominantly female environment, were often both less able and less willing to cope.

So the difficulties ex-Society members faced within the new Institute can be explained and clarified in a number of ways, using theories of group interaction. But how far were the differences exacerbated because the ex-Society members were women and most of the Institute members men?

REASONS FOR DECLINE IN WOMEN'S MEMBERSHIP: STEREOTYPING AND SEX DIFFERENCES

The issue of gender relationships and sex stereotyping has not been much explored in the general literature about groups (though Tajfel, 1981: 341 provides a brief discussion linking the two approaches). On the whole it has been feminist theorists and psychologists interested in sex differences who have explored the effects of sex stereotyping (using Bouchier's definition of feminism as 'any form of opposition to any form of social, personal or economic discrimination which women suffer because of their sex': 1983: 2). It should be noted that 'Sociologists make an important distinction between "sex" and "gender". The term "sex" refers to the biological difference between males and females, while "gender" refers to the socially determined personal and psychological characteristics associated with being male or female' (Garrett, 1987: vii). This differentiation is useful and is followed where appropriate, but many writers, especially in the past, have used the word 'sex' for both, and this is reflected in quotations and comments.

The existence and content of stereotypes

In Allport's words, 'a stereotype is an exaggerated belief associated with a category. Its function is to justify (rationalise) our conduct in relation to that category' (Allport, 1958: 187). Much of the early work on stereotyping focused on race (for example, Adorno *et al.*, 1950). Nevertheless, the concept of stereotyping was seen as important in relation to the role of women in society and has been the focus of much psychological research. Maccoby and Jacklin in 1975 carried out a major review of this literature. Their breakdown of the kinds of attribute which tend to be ascribed to males and females gives a useful guide to some of the contents of stereotypes and is shown in Figure 8.2. Tajfel (1981) provides a useful summary of the social and cognitive processes involved in the formation of stereotypes. Feminist writers have consistently stressed the discriminatory importance of the way in which male and female roles or characteristics are seen as being predetermined (for example, Figes, 1970; Firestone, 1980; Mitchell, 1971; Friedan, 1975). Socialist and radical feminists tend to put more stress on the role of economic, social and cultural exploitation of women than on psychological processes *per se* (for example, Mitchell, 1971; Firestone, 1980).

Wolpe (1977) gives a good simple expression of some common aspects of sex stereotype roles as expressed in daily life:

> In regard to the division of labour within the family, it is the woman, in her role as housewife and mother, who is predominantly responsible for child-minding, housework, and care of the family. In this familial role, women not only reproduce the labour force through the birth and care of children till they reach an age when they can enter the labour market, but they also provide domestic services and 'products' (meals, making and care of clothes, etc.) which enables those members who work to reappear daily in their jobs.
>
> There is also a sexual division of labour within the economic structure of the society outside of the family . . . women are relegated to what has been termed a secondary labour market and this . . . is not unconnected with the division of labour in the family.
>
> (Wolpe, 1977: 1)

Adams and Laurikietis (1976: 19) give another simplified expression of stereotyped views of gender roles as they affect young people. They talk about various myths about men and women:

Unfounded belief about sex differences

---

1 That girls are more social than boys.
2 That girls are more 'suggestible' than boys.
3 That girls have lower self-esteem.
4 That girls are better at rote learning and simple
   repetitive tasks, boys at tasks that require
   higher-level cognitive processing and inhibition
   of previously learned responses.
5 That boys are more 'analytic'.
6 That girls are more affected by heredity, boys by environment.
7 That girls lack achievement motivation.
8 That girls are auditory, boys visual.

Sex differences that are fairly well established

---

1 That girls have greater verbal ability than boys.
2 That boys excel in visual-spatial ability.
3 That boys excel in mathematical ability.
4 That males are more aggressive.

Open questions: too little evidence or findings ambiguous

---

1 Tactile sensitivity.
2 Fear, timidity and anxiety.
3 Activity level.
4 Competitiveness.
5 Dominance.
6 Compliance.
7 Nurturance and 'maternal' behaviour
   (including passivity).

Figure 8.2   Maccoby and Jacklin: beliefs about sex differences
Source: Summary extracted from Maccoby and Jacklin (1975: 349–54).
(Note: For criticisms of Maccoby and Jacklin's methodology or summarisation, e.g. see
Block 1976; Griffiths and Saraga 1979.)

Myth number one:
   Females are passive and unaggressive.
   They care for and support others.
   They are domestic and dependent.
   They are easily upset and emotional, given to crying.
   They should be dominated by men.

Myth number two:
   Males are active and aggressive, independent and adventurous.
   They can cope with the world.

They are logical and unemotional.
They ought to be able to dominate women.
They are tough, violent, ambitious, ruthless.

Myth number three
If males don't behave in a masculine way and females in a feminine way, there is something wrong with them.

The nature and origins of psychological differences between men and women

A great deal of the literature about stereotyping centres around the existence and origins of psychological differences between men and women. Such theoretical discussions do have practical outcomes because they influence views on what should be done about inequality. Maccoby and Jacklin (1975: 360ff.) carried out an extensive review of this literature and discuss

> three kinds of factors that affect the development of sex factors: genetic factors, shaping of boylike and girllike behaviour by parents and other socializing agents, and the child's spontaneous learning of behaviour for his sex through initiation.

In their review of the evidence, summarised in Figure 8.2, they identify a small number of factors where they consider that there is some evidence for innate sex differences. However, other writers have criticised Maccoby and Jacklin's findings as being incomplete, misrepresenting the evidence or failing to take account of some factors (for example, Block, 1976). Griffiths and Saraga (1979) continue this criticism with a good analysis of some of the biases inherent in much of the research and argue for a different framework for analysis. The essential point of this argument is that explanations based on genetic factors, in their view, justify and defend the current status quo. Griffiths and Saraga point out the need to take a wider perspective than many psychologists do. The study of socialisation into sex roles is incomplete if it neglects the origins of these roles which, in Griffiths and Saraga's view 'need to be examined and understood as a product of women's oppression under capitalism' (1979: 36). Garrett (1987) provides an overview of the different theories and of the feminist critiques of them. Connell (1987) makes a comprehensive analysis of the weakness of sex role theory and the limitations of psychological 'sex difference research' and its methodology. He concludes:

Recent research has *not* shown that Maccoby and Jacklin systematically underestimated sex differences. A striking conclusion emerges. The logic of the genre focuses on difference and its explanation. In fact the main finding from about eighty years of research is a massive psychological similarity between men and women in the populations studied by psychologists.

Connell argues convincingly that it is only the social investment in the notion of sexual character which serves to perpetuate it. However, in the period under discussion it was still very influential in popular thought. Even in recent years popular theories about whether sex differences are innate or not heavily influence attitudes to programmes of action for equal opportunities.

In examining sex stereotyping further, the focus will be on those aspects which might have a significant effect on the interaction of men and women in the Institute after unification or on women's careers in housing generally.

Differences in generalised intellectual abilities

'It appears that the work of males is more valued even if it is identical to that of females' (Shibley Hyde and Rosenberg, 1976: 100). Goldberg's famous study showed that, female college students gave lower grades to essays which were given the names of female authors (Goldberg, 1968).

The evidence from interviews with Institute and Society members indicates that this aspect of stereotyping was not the most influential. Because of the different procedures involved in selection and training, both Society and Institute interviewees perceived Society members as having a higher general standard of education and ability. If this stereotyping was operating in housing, it was more likely to have affected women in housing lower down the career ladder, or in competition for senior posts.

Aptitude for specific types of study or occupation

In Maccoby and Jacklin's categories this is expressed, for example, in the items 'that girls have greater verbal ability than boys', 'that boys excel in visual-spatial ability', 'that boys excel in mathematical ability'.

Supposed differences in aptitude for different types of study have frequently been used as part of the justification for women not playing a significant part in particular occupations. Much of the argument has centred around the general area of science (for example Weinreich-Haste, 1979). The technical, building side of housing has been male dominated, so perceived weakness here might also have affected the status of women in mixed-sex interaction in the new professional Institute (this is still an important issue). Evidence on genetic predisposition in this respect is still controversial: 'Great uncertainty exists in this area, and it is dangerous to assume at the present time that there are such things as clearly established sex differences in intellectual functioning' (Murphy, 1979: 161).

Shibley Hyde argues that 'Females consistently show poorer spatial and mechanical abilities than males do' (Shibley Hyde and Rosenberg, 1976: 84). She also seems to give some credence to the view that this difference in ability is governed by a sex-linked gene though it may be reinforced by social and psychological factors (ibid.: 85). Other writers (for example Sayers, 1979) emphasise the effect of sex role pressures. However, Shibley Hyde goes on to demonstrate how unrealistic it is to argue that differences in occupations are due to genetically determined spatial ability even if this exists.

> The current estimate of the frequency of the recessive sex-linked gene for high ability at spatial visualising is 0.5 (Bock and Kolakowski, 1973). Hence, we would expect 50 percent of all men to have relatively high spatial ability, as compared with . . . 25 percent of all women. If this genetically determined ability were the sole determinant of becoming an engineer, we would expect to find the ratio of men to women in this profession to be . . . 67 percent men and 33 percent women. Clearly the 1 percent of engineers who are women is far from what we would expect if sex differences in spatial ability were alone responsible for women's lesser participation in the profession. The difference must be attributed to various cultural sources.
>
> (Shibley Hyde and Rosenberg, 1976: 85)

What Shibley Hyde says of engineering must also be true of those occupations associated with the building industry where spatial ability plays a part – for example planning, architecture and the technical side of housing work. Connell's (1987) arguments similarly undermine the logical basis of much of this research.

Whatever the justification for the arguments about different aptitudes, we have seen that the Society stood in a rather ambivalent position here. Octavia Hill had emphasised the need for a basic businesslike approach which would include the ability to deal with figures and the technical side of housing. This had been continued by the Society in its training. To this extent Society members could be seen as overcoming this stereotype, and many did. In addition, some of the senior members had full RICS qualification which, because of its heavy 'technical' content, tended to compensate for the disadvantage. However the Society had at times used the stereotype of the 'caring' role attached to women and this was likely to be seen as the antithesis to technical/financial and managerial role.

Mitchell points out: 'Bearing children, bringing them up, and maintaining the home – these form the core of woman's natural vocation, in this ideology . . . The causal chain then goes: maternity, family, absence from production and public life, sexual inequality' (Mitchell, 1971: 106, 107). In addition, by an extension of this role, women are seen as possessing the psychological attributes associated with caring and as being responsible for the caring and nurturing role within society generally.

Because of the tendency of psychological characteristics to be seen as bipolar, to some extent this nurturing, caring, female role has been viewed as the counterpoint to the active dominant technical role of men. (See, for example, Sayers' discussion of psychological sex differences in Hartnett *et al.*, 1979: 46–55.) There is extensive evidence that the main burden of caring for the sick, disabled and elderly does fall on women. Concern about this has recently led to the formation of the Association of Carers (Toynbee, 1984).

The image of women having this 'caring, welfare, female' role had appeared in housing in the past, was to some extent seen as the opposite to 'business management' and the 'technical' side, and could have affected the interaction between male and female members of the Institute. Those members most concerned with unification did on the whole perceive each other much more realistically. But women members of the committee concerned with unification grew heartily sick of having the amalgamation referred to jokingly as a 'marriage' – perhaps a good illustration of the way in which jokes can be used to belittle women. The practical reality of having to care for children or other dependants, in particular elderly relatives, could, and did, hold women back in their careers.

This stereotype resurfaced in a rather interesting way in the Institute in 1984. An article purportedly reviewing the history of the amalgamated Institute on its long awaited award of a Royal Charter, 'The long road to royal recognition', was published in the

December 1984 issue of the Institute's journal *Housing*. The article reflected an orientation which proved to be offensive both to women ex-Society members and at least one of the men. It included the view that 'The Octavia Hill tradition of women serving a primarily caring role on the charitable side of the housing profession was no doubt intensified when it was the male-dominated Institute who first established housing management as a recognised career in the local authority world' (Hirsch, 1984: 4). Although the writer had consulted various sources, the views reflected seemed largely to be those of a senior male member of the Institute, John Macey.

The article provoked a spate of letters. One typically referred to the 'disgustedly patronising tone of the remarks about the former Society of Housing Managers' and, in relation to the quotation about the 'Octavia Hill tradition', commented, 'The whole basis of Octavia Hill's approach was efficiency . . . the people she trained were taught to run an efficient and human service, while never losing sight of the broader issues (contrary to John Macey's very typical remark quoted later in the article)' (Houstoun, 1985). The remark referred to was: 'The larger authorities have tended to appoint men, who have had a wider view of what housing management means than women' (Hirsch, 1984: 4).

If this stereotype could surface so strongly 30 years after the unification of the Society and the Institute there certainly seems to be ground for supposing that it would have been influential in the 1960s.

Dominance and leadership

The same kind of stereotyped thinking, which sees women as naturally having nurturing, social and submissive characteristics, sees men as being more aggressive, self-confident, ambitious and naturally having a dominant and leadership role (for example, Maccoby and Jacklin, 1975: 353–354; Sayers, 1979: 51–55). This leads to problems for women who wish to take on leadership or creative roles (Reid and Wormald, 1972: 161, 162; Hargreaves, 1979: 185–199).

This factor could have been important in the interaction of Society and Institute members. If the assumption was that men would take on the leadership role then the failure of women to survive on Council and in senior positions within the Institute is more explicable.

From interviews with the women members, it is clear that the majority did not overtly subscribe to this view. Indeed, they recognised that there were women in the Society capable of carrying out these tasks and would sometimes say that they were often more capable than men. However in actual social situations, women who do not subscribe to these stereotypes may still be influenced by them.

From interviews with the smaller number of male members involved in unification, it was clear that most of them appreciated the abilities of the senior women in the Society. They did also imply, however, that Society members were not so well known in the Institute and that attitudes of members of the negotiating team were different from those of the Institute members. There was a strong association between 'success' in the Institute and holding down the big jobs. Perception of women as not having leadership qualities has been indicated as important in affecting some women's progress in management jobs, particularly when organisations get large (for example, Cooper and Davidson, 1982: 365, 33). This factor is so pervasive that it seems likely that directly or indirectly the perception of women as not having a natural leadership role did affect the women in the new Institute.

It would also affect interaction in meetings. 'Women in virtually every group in the United States, Canada and Europe soon discovered that, when men were present, the traditional sex roles reasserted themselves regardless of the good intentions of participants. Men inevitably dominated the discussions' (Freeman, 1979a: 563). This quotation, describing the early days of the women's liberation movement, summarises much of the feminist argument in this respect. Men's dominance in society is reflected in their dominance in meetings. Freeman was explaining why women in the women's movement soon found it necessary to restrict their meetings to women. Women-only activities have remained a central feature of the women's movement though not without continuing discussion (for example Mitchell, 1971: 56–58).

It is likely that male dominance at meetings would reinforce difficulties Society members had in fitting in both to branch meetings and committees. It seems logical to argue that this difficulty would exacerbate the feeling of Society members being 'deviant' and 'rejected' which have been discussed in the context of relationships within groups.

The members of the Society most concerned with unification were so anxious to make it work that they rejected any moves to continue some form of separatist organisation or network (interviewees). It is interesting to speculate whether any organisation of this type might have safeguarded women's interests better within the Institute; it has certainly proved necessary in order to revive them (see Chapter 11).

Men's dominance in language

Dale Spender (1980), Robin Lakoff (1975), and others have shown how men's dominance is enshrined even in the conventions of language and discussion. Detailed findings on the conventions of conversations between men and women indicate how easy it is for men to dominate such discussions and how women are seen as 'pushy' or aggressive or shrill when they try to assert themselves. It is easy in mixed sex meetings for men to dominate. Dale Spender believes that these theories are easy to demonstrate in any mixed group and it is a continual struggle for women 'to get in the 50 percent worth'.

> Indeed, there are indications that when women do try and speak, and interrupt at the same rate as men in a mixed group, they are often labelled as persistent, tenacious and annoying by the male participants.
>
> (Cooper and Davidson, 1982: 43)

Such factors were likely to have been operating within the Institute both at Council and at branch level. In such circumstances, for example, the same weight is not given to a female contribution to discussion as to a male one (Spender, 1980: 47). This factor is very insidious and extremely difficult to combat because it is covert. It could still be operating within the Institute today. Certainly some of the women who fought for women's rights within the Institute in the 1980s felt that they were labelled as 'persistent, tenacious and annoying' by the male participants.

Men's dominance in language is likely to have been important in reducing both the amount and influence of women's participation at branch and committee meetings.

Male and female styles of organisation

The preferred style of women operating in organisations has received some study of recent years, with an emphasis on the importance of informal networks for women. Estelle Freedman has studied 'female institution building' in the nineteenth century and argues that

> When women tried to assimilate into male-dominated institutions, without securing feminist social, economic, or political bases, they lost the momentum and the networks which had made the suffrage movement possible.
>
> (Freedman, 1979: 524, 525)

She argues that today it is equally important for women within mixed institutions to create female interest groups and support systems; otherwise they will be co-opted into traditionally deferential roles or assimilated. While women need to move into the public field dominated by men, they need also to affirm the value of their own culture.

Ryan, in her study of some early American institutions, claims that the associations which women set up exhibited a new mode of organising:

> Most were congregations of peers. . . . Most rejected a rigid governing hierarchy and condescending manners. . . . All these associations occupied a distinctive space in the social order of the community, somewhere along a muted boundary between private and public life. . . . The association relied on informal but expansive social ties, a voluntary network of like-minded individuals. . . . Social organisations of this nature are particularly receptive to female participation.
>
> (Ryan, 1979: 68,69)

As far as more recent organisations are concerned, numerous writers have commented on the desire of women's groups to use non-hierarchical modes of organisation (for example, Coote and Campbell, 1982; Bouchier, 1983; Rendel, 1981). Some writers see the desire for non-hierarchical organisation as a barrier to the women's movement achieving the redistribution of power which is necessary to its ends (Bouchier, 1983; Rendel, 1981).

The Society of Housing Managers, as we have seen, did not reject hierarchical organisation. But interviewees claimed there was a conscious effort to make this broad based and flexible and to 'bring on' younger members. Formality in the meetings was kept to a very practical level. 'Informal and expansive social ties' were a marked characteristic of the Society. The comments of Society members quoted earlier in this chapter indicate clearly that the style of the Institute meetings and proceedings was very different from that of the Society, with a greater degree of formality in the new Institute and 'dangling of badges and chains of office'. There also seemed to be a greater emphasis on hierarchy. The old predominantly male Institute had, like most organisations, served a social role but comments from male interviewees indicated that involvement was rather 'cooler', more impersonal and specifically related to the job. It could be argued that, in entering the new Institute, women not only lost their existing social support system, but also were put into a form of organisation which was less congenial to them. Those women who had been used to working in male-dominated organisations were probably better prepared to cope with the new situation. Nevertheless, it is important to recognise that all these factors working together would mean that women were handicapped in getting their views across, more likely to be defeated in committee and to be in the position of the deviant described earlier. Many 'shrugged and turned away'. Women could keep up formal membership of the Institute as a professional body without having to face the hazards of being an increasingly uncomfortable minority in meetings. This explains why women's representation on Council fell more sharply than their membership of the new Institute.

CONCLUSIONS

The discussion in this chapter has concentrated on changes within the Institute of Housing, and the reasons for them. Two main bodies of theory, group interaction and stereotyping, have been used to comment on these changes.

Between them they provide many powerful explanations of women's reduced participation in the Institute after 1965. In particular, the assumption of men's 'natural' leadership and dominance and the way this is expressed in language and in meetings seem to have been very influential. These reduced the perceived influence of women by reducing participation by women on Council and at branch meetings. A 'knock-on' effect would then result, making other women less willing to participate in any active way.

But there has been evidence, both from the membership statistics and from interviews, that women may also have been affected by changes in housing employment generally, and that these changes are complex rather than simple. Patterns of employment must be considered to identify the influence of external events as well as interaction within the Institute.

9 Women in housing employment, 1965 to the 1980s

THE HOUSING BACKGROUND

Housing policy

Housing policy in this period is complex but certain major trends were especially significant for housing management and administration. By 1965 the Conservatives were limiting local authority house building to slum clearance needs. They argued that housing subsidies should be concentrated on those in greatest need by raising rents and providing personal subsidies. A policy of encouraging housing associations to form a third arm of housing provision was beginning to receive wide support. Although the Labour party instituted a review of housing finance when in power in 1977 it had come to accept owner occupation as the 'natural form of tenure', while the Conservatives continued their strong support for owner occupation. This changing emphasis was bound to affect the public sector. Both parties faced problems about the enormous cost of replacing substandard housing. The switch from slum clearance to improvement of older housing, marked by the 1969 Housing Act, was to some extent bipartisan policy. By the 1980s the Conservatives were actively 'encouraging' local authority tenants to change sector via the Right to Buy.

The influence of housing policy on administration over this period arose not just from the content of policies but from the speed and abruptness of change and the contradictions. Political factors meant that major administrative schemes, such as the national rent rebate and rent allowance scheme initiated in 1972, were brought in with insufficient consultation and preparation time and insufficient resources in the local authorities to deal with them. The chaos which resulted in some places rebounded on the local authorities rather than on Central Government. Similar problems were repeated, for example in the 1980s with improvement grants, and continue into the 1990s. The balance of power between Central Government and local was shifting, but the appearance of local democracy continued to be preserved, increasing the strains for those employed at local government level. Political allegiances started to impinge more sharply on chief and senior housing officers (Slizowski, 1984).

Housing administration, 1960 to the 1980s

Unified housing management

In 1959 the Minister of Housing and Local Government, addressing the Institute of Housing Conference, argued that the fragmented approach, with housing functions split between several departments, did not suit 'housing management on the large scale as we know it today' (Kemp and Williams, 1991: 135). This marked a change of tack for Central Government which had hitherto adopted a permissive approach to local organisation of housing functions. Was it part of the trade-off in the pressure to form a unified professional body? Or were both part of a realisation by Central Government that housing management needed to be improved in order to meet changing circumstances? Possibly the latter, since all subsequent reports until the 1990s stressed the need to create unified housing departments. Most authorities did create 'unified' housing departments but functions remained split in a number of authorities and were subject to continuing change (see, for example, Housing Research Group, 1981).

The comprehensive housing service

Not only were departments now to be unified but their functions were to be changed. The Seebohm Report in 1968 argued for the development of a comprehensive housing service in which local authorities should take the broadest view of their responsibilities and be concerned with housing in all sectors and with differing ways of assisting people in housing difficulties. This argument was strengthened with a growing reliance on rehabilitation policy in the 1960s. Housing authorities increasingly had to become involved with rehabilitation of the private and owner-occupied housing. The old concentration on the building and management of local authority estates was seen to be no longer the answer. The Cullingworth Committee (MHLG, 1969) endorsed this view, which was taken up by Central Government. London boroughs like Lambeth began to introduce the ideas of housing aid centres and a more comprehensive approach (Laffin, 1986: 91). Although the idea of the comprehensive housing service was later criticised on the grounds that local authorities are unable to influence many aspects of the housing market (Harloe, Issacharoff and Minns, 1974; Cockburn 1977), it remained into the 1980s the formal pattern to which housing departments were meant to conform.

The effects of local government reorganisation

Local government reform in London in 1965 and in the rest of the country in 1974 meant that the size of departments multiplied by three or four times. Given that the administration had often been under strain on the old departments, this increase in size produced severe problems in the delivery of services. Growing size also led to increasing pressures for efficiency (see, for example, Laffin, 1986: 108–109 for an account of this period). Interviewees frequently commented unfavourably on the effect of local government reorganisation on the efficiency and quality of service to the public. Many had argued for the need to keep an effective, personalised management close to the ground (since this is what they saw as the Octavia Hill tradition). But they were told that their views were old-fashioned, and that this detailed approach was not needed in the 1970s. Kemp and Williams (1991) comment:

> throughout the 1960s and 1970s local authorities moved further and further away from close, personal contact with their tenants, partly in attempting to achieve economies of scale. Social and physical distance crept into the municipal landlord/tenant relationship, making housing management more extensive, more remote and increasingly depersonalised.

Power has pointed out that a number of changes which occurred over the post-war period in housing management ultimately disadvantaged the whole service. She argued that both direct management and maintenance subsidies and unitary management organisations could have made the job of public sector housing management a great deal easier (Power, 1987: 90).

The effects of slum clearance

Slum clearance peaked in the 1960s and was being wound down by the 1970s but a number of critiques of this policy and of housing departments' part in it were published in the 1970s and 1980s (for example, Coates and Silburn, 1980; Muchnick, 1970). The odium of housing departments' involvement in what was often seen as an unpopular policy still clung and in some authorities the battle of clearance versus redevelopment was still being fought. For many authorities Central Government policy and commercial pressures ensured a legacy of high-rise blocks, which later became the un-popular face of council housing (Dunleavy, 1981). Impersonal large-scale management of people was one of the issues attacked in many of the critiques (for example, Lambert, Paris and Blackaby, 1978).

Growth in housing associations

Although housing associations accounted for only just over 2 per cent of the total stock of dwellings they became increasingly important in the 1970s both in an accelerated rate of new build and in rehabilita-tion work (Balchin, 1981: 139). Because associations were beginning to receive more help from public money, the government stepped in to try to improve standards of management. After the 1974 Housing Act associations had to be registered with the Housing Corporation to receive public money, but more generous grants were also available which led to a rapid expansion (Short, 1982: 191,192). Although there were over 2,000 housing associations in the United Kingdom by 1979 only 20 per cent of those had paid staff full-time. Only 30 owned more than 1,000 dwellings (Balchin, 1981: 140). The period of increased activity for housing associations meant new opportunities for staff and higher salaries (Allen, 1981: 123–181).

Residualisation

In 1977 the Housing (Homeless Persons) Act made rehousing certain categories of homeless people obligatory. Increasingly the emphasis was on housing on the basis of 'need' rather than, for example, time on the waiting list. At the same time the supply of social housing was gradually diminishing because of government policy. The effects of this on housing management were given little attention at the time. This meant there was growing dislocation between the official view that council tenants were getting more prosperous and more educated and needed less intensive housing management and the reality for some inner city authorities that they were rehousing poorer tenants often in high-rise blocks that were more difficult to manage (Housing Research Group, 1981: 34, 35). By the 1980s many commentators forecast that the effect of sales of council housing under the Right to Buy and other Conservative government policies would be to condemn local authorities to a residual role and to emphasise the importance of welfare in management (for example Murie, 1985; English, 1982; Park, 1984). Contradictions between this residualised role and government desire for 'business management' were also becoming marked by the 1980s.

Quality and equality issues

Shelter's publications, for example *Homes fit for heroes* (Griffiths, 1975) criticised council housing management on many grounds – lack of rights, lack of choice in allocations, restrictive tenancy conditions. In addition, housing departments were often accused of discriminatory and judgmental attitudes to the homeless (see, for example, Greve *et al.*, 1971; Glastonbury, 1971). Efficiency in such services as repairs was also under attack (NCC, 1979). A series of reports, beginning with the Runnymede Trust's *Race and council housing in London* (1975), were illustrating not only the disadvantaged housing position of black minorities but also the existence of indirect and direct discrimination in housing departments and associations.

GENERAL TRENDS IN WOMEN'S EMPLOYMENT

Women's participation in the labour force, 1965–80

The long-term trend of increased female participation continued. The contribution of single women, who had been the major part of the workforce before the war, declined, being replaced by greater participation of married women in the workforce (Joseph, 1983: 68–77). Joseph stresses two main factors that led to more female participation in the labour force over this period: it may have been easier for women to combine work and a family because of wider availability of part-time work or the existence of more labour-saving devices; more women may have wanted to work because of changing aspirations to lead a fuller life, a desire to earn more money to attain a higher standard of living. Martin and Roberts' work gives prominence to similar factors. 'In fact almost all the growth in employment from the 1950s onwards can be attributed to the increase in part-time work' (Martin and Roberts, 1984: 1).

There was some modification of the trend by the end of the 1970s, when there was a decline in the participation rates of older men and women. 'Until 1977, labour force participation rates for married women rose steadily from 42 per cent in 1971 to 50 per cent in 1977. . . . But since 1977, the increase stopped and there is even some evidence of a slight fall back in these rates' (Joseph, 1983: 119).

Other writers have stressed that the way in which unemployment statistics are calculated underestimates women's unemployment. This is important because women, especially those working part-time, may be the first to be turned off when the available jobs are shrinking (Joshi, 1984). In addition, by the 1970s, writers were pointing out the extent to which the advent of new technology might disproportionately reduce women's jobs (Bruegel, 1979). Walker (1988) has analysed the way in which the welfare cuts of Thatcher governments made it more difficult for women to work while the government was trying to expand women's share of the labour market and families were increasingly dependent on dual incomes. This meant, once again, that movements in women's employment were complex rather than simple. In the early 1980s, women's employment did not reduce more than men's, though the shift to part-time continued. (Changes in the calculation of unemployment statistics also make comparisons difficult.)

Women's employment patterns

The M-shaped pattern and part-time work

As Martin and Roberts (1984: 1) note, by 1979, 61 per cent of women of working age had a job, compared with 54 per cent in 1971. Most of this rise was accounted for by increasing numbers of women returning to work after having children and taking a shorter time out of the labour market. This gave rise to a bimodal career pattern (producing an M shape when graphed) common to women in this country and the USA. 'This has the effect of polarising the female workforce into two age groups – the young and the returners' (Silverstone and Ward, 1980: 10).

However, Rubery (1988) demonstrates that 'In France and the US the M shape has now virtually disappeared giving a relatively flat pattern of participation over prime age ranges. . . . The Italian participation rate pattern seems to be adjusting from the traditional pattern to the flatter profile without the intermediate stage of an M shaped pattern'. She concludes that the diverse pattern of participation rates can be related to differences in the pattern of child care and in the sources of income available to females – once again a reminder of the way in which the context influences the pattern of disadvantage (Rubery, 1988: 276).

So the fact that women participate in the labour force does not mean that women compete on equal terms in the labour market. The impact of domestic labour and of childbearing causes many to choose part-time jobs which channel them into low-paid work. Even women who have had professional or managerial jobs, and are often better placed for returning, find they are at a disadvantage. They are faced with conflicts over child care and the way that their role is seen in society (Silverstone and Ward, 1980: 17). They may find it difficult to return to jobs at their original level or need retraining (Povall and Hastings, 1982: 1). If they have been absent from employment in their late twenties and early thirties they find that this is a crucial period for promotion and they have missed the boat (Fogarty, Allan and Walters, 1981: 250). The bimodal pattern did not fit into the expected career paths in large organisations, which were still too often based on male career patterns (see, for example, Ashridge Management College, 1980: 134, 135; Fogarty, Allen and Walters, 1981: 9, 10).

However, there is evidence by the 1990s that more women are remaining in employment when they have children. Fear of losing jobs is one motivator but better maternity and child care provision from the employer also plays a part.

Occupational segregation

It is now well established that it is necessary to look not only at the extent of women's participation in employment but also at the types of jobs they are in. Occupational segregation – the extent to which jobs or types of jobs are allocated predominantly to one sex – is important. Hakim (1979) distinguishes two kinds of occupational segregation – horizontal and vertical.

> Horizontal occupational segregation exists when men and women are most commonly working in different types of occupation. Vertical occupational segregation exists when men are most commonly working in higher grade occupations and women are most

commonly working in lower grade occupations or vice versa. The two are logically separate.

(Hakim, 1979: 19)

Hakim's data confirms the persistence of vertical occupational segregation nationally:

Within each occupational group, women tend to be over-represented in the less skilled, lower-status or lower-paid jobs, while men are over-represented in the highly skilled and managerial jobs.

(Hakim, 1979: 31)

This is one aspect of women's employment which has been well researched. Hakim also points out that changes have often been in the direction of greater segregation rather than integration of sexes in the work sphere. For example, about three-quarters of all clerical workers were women in 1971 compared to only 21 per cent in 1911. The proportion of women in managerial and administrative positions or in lower professional and technical occupations actually declined between 1911 and 1961, though 1971 figures suggested that women were regaining some of this ground. Although some improvements occur, changes have tended to cancel themselves out with the result that 'Overall, there has been no change in the degree of occupational concentration, and no change in the degree of occupational segregation, since the turn of the century' (Hakim, 1979: 29).

Martin and Roberts were critical of applying standard occupational classification schemes to women. They worked out their own scheme and used it to compare socio-economic groups of men and women working over a period of time (Martin and Roberts, 1984: 20–33). Their data also indicated overall stability in occupational segregation 1965–80.

The most important point for this study is that there is no evidence of a general decline in women's share of professional jobs between 1965 and 1977 which would explain the decline in women's participation in the Institute over that period. There was some decline in women's share of managerial jobs up to 1961 and the M-shaped career pattern may also have had an effect. It is logical therefore to look for factors specific to housing which might explain the very marked changes identified earlier.

WOMEN'S EMPLOYMENT IN HOUSING 1965–84

Chief Housing Officers in London

Table 9.1 looks at the distribution of men and women Chief Housing Officers in the London region from 1955 to 1984. The table shows very clearly that the number of women Chief Housing Officers in London was already dropping prior to 1964: 47 per cent women in 1955 had fallen to 22 per cent women in 1964. But the drop is even more abrupt between 1964 and 1966 (down to 8 per cent in 1966), illustrating the importance of local government reorganisation. By 1984 there was only one woman chief officer left.

These data illustrate a number of points. They tend to confirm the interviewees' comments that most of the strength of the Society was in London, which had quite a respectable share of the higher posts in 1955. All except one of these women managers in 1955 were members of the Society. But 1955 probably represents the culmination of this trend. By 1964, the number had already dropped. Moreover all five of the women in post in 1964 had been in post in 1955 – no new appointment of a woman manager had been made. The London data suggest that the changes going on in housing and local government had already begun to disadvantage women prior to 1964 and that local government reorganisation accelerated this trend.

Chief Housing Officers outside London

Problems with the data only permit analysis of the municipal corporations. Table 9.2 illustrates that women Chief Officers were always rarer outside London. Reorganisation into larger authorities seems to

Table 9.1 Men and women chief housing officers in London* 1955, 1964, 1966 and 1984

Year	Met./London boroughs listed as having Chief Housing Officer	Men	Women	% Women
1955	27	19	9	47
1964	28	23	5	22
1966	28	26	2	8
1984	34	33	1	3

Source: Municipal Year Books for 1955, 1964, 1966 and 1984
* London here means: prior to 1965, the Metropolitan boroughs and the LCC; after 1965, the London boroughs, the GLC and the City of London Corporation.

Table 9.2 Men and women chief housing officers

Year	Total listed	Men	Women	% Women
Municipal corporations				
1955	337	306	31	9.2
District councils				
1976	–	–	–	4.0
1984	320	316	9	2.8

Sources: 1955 and 1984 data: Municipal Year Books, 1955 and 1984. (Only those authorities listed where it was possible to identify gender of postholder have been included.) 1976 data: Brion and Tinker (1980)

have accentuated this.

The NFHA Council

It is not possible to get any usable list of chief officers of housing associations until the advent of registration with the Housing Corporation in 1974. However, the record of membership of the Council of the National Federation of Housing Societies does give some evidence about the standing of women in the housing association world. The NFHA was

> formed in 1935 partly at the request of the (then) Minister of Health, to take over the work previously carried out by the Garden Cities and Town Planning Association, and to co-ordinate and make representations on behalf of the 35 Societies then in existence and affiliated to it.
>
> (NFHA, 1958)

It remains the officially recognised body representing housing associations.

Table 9.3 shows once again a relative decline in the influence of women in the 1970s. While women were never very heavily represented on the NFHA Council, in the 1930s they usually formed about a quarter of the membership, and at certain times – the 1940s and 1950s – up to a third. As the table shows, the proportion of Council members who were women began to fall in the 1960s, with a more marked fall in the early 1970s; the lowest point was reached in 1974, with no women members of Council. This has a close correspondence with the movements in the proportion of women on the Institute of

Housing Council analysed in the last chapter. Again, there is some revival by the early 1980s.

1976: The City University staff study

This survey was part of the Education and Training for Housing Work Project carried out at The City University between 1975 and 1977. The Staff Study (Education and Training for Housing Work Project, 1977) was carried out in 12 organisations, 6 housing departments and 6 housing associations, chosen to represent different types of authority and geographical area.

Grades of men and women in housing organisations

The position of men and women was analysed in terms of local authority grades, and housing association employees were amalgamated into appropriate grades by salary. Figure 9.1 shows the distribution by grade of men and women in the Staff Study sample. The most important fact is that although there were many women employed in these organisations they were generally at lower grades than men (Brion and Tinker, 1980: 49).

Types of work done by men and women

There were differences in the type of work done by men and women, between different types of organisation and different functions.

For example, women constituted 15 per cent of housing management section heads and only 7 per cent of the section heads in the development and rehabilitation function (the function most closely linked with the building industry) (Brion and Tinker, 1980: 51). Thus the 'Staff Study' identified both vertical and horizontal occupational segregation in housing.

Educational qualifications and grade

Another striking finding of the survey was in regard to the relationship between educational qualifications and grade, though the sample numbers were small here. Of the 31 'qualified' men in the sample, 22 had reached grades of principal officer and above. But only 8 out of the 18 qualified women had reached these grades, indicating a significantly lower level of achievement in terms of work grade (Education and Training for Housing Work Project, 1977: 52).

Table 9.3 Men and women members of the Council of the NFHA 1936–84

Year	Total	Men	Women	% Women
1936	18	14	4	22
1937	17	12	5	29
1938	20	15	5	25
1939	21	15	6	29
1940	21	15	6	29
1941	21	15	6	29
1942	21	15	6	29
1943	21	15	6	29
1944	21	15	6	29
1945	20	14	6	30
1946	18	13	5	28
1947	20	16	4	20
1948	20	16	4	20
1949	20	16	4	20
1950	20	16	4	20
1951	19	14	5	26
1952	20	14	6	30
1953	20	14	6	30
1954	20	14	6	30
1955	21	14	7	33
1956	21	15	6	29
1957	21	15	6	29
1958	19	13	6	32
1959	19	13	6	32
1960	21	14	7	33
1961	21	16	5	24
1962	23	19	4	17
1963	24	20	4	17
1964	25	21	4	16
1965	25	21	4	16
1966	25	21	4	16
1967	23	18	5	22
1968	23	18	5	21
1969	25	21	4	16
1970	23	20	3	13
1971	21	20	1	4
1972	23	22	1	4
1973	20	19	1	5
1974	21	21	0	0
1975	22	21	1	4
1976	21	19	2	9

Year	Total	Men	Women	% Women
1977	21	19	2	9
1978	23	20	3	8
1979	24	22	2	8
1980	26	24	2	8
1981	30	26	4	22
1982	32	27	5	17
1983	37	32	5	14
1984	30	25	5	17

Source: Copies of Council membership in Annual Reports supplied by NFHA.
Members and chairmen have been counted but not presidents and vice-presidents.

Women's employment in 1984: the NFHA study

The NFHA survey was carried out in the first half of 1984 by a postal
questionnaire to all NFHA member associations which employed five
or more full-time staff. Out of 188 associations 137 usable replies
were received (73 per cent response rate). At the same time the
NFHA carried out a review of the career patterns of a sample of
staff who had attended NFHA seminars and a review of current
recruitment practice by following up a selection of advertisements
for housing association jobs. A major report with practical recommendations was subsequently published (NFHA, 1985). The tables in
the following section have been recalculated from raw data since
the information was not usually shown in this form in the NFHA
publications.

Full- and part-time employment in housing associations

Figure 9.2 shows that there were a total of 7,146 employees, of 4,675
(65 per cent) of whom were full-time and 2,471 (35 per cent) part-
time. Housing associations thus employed a substantial number of
part-time staff and, as the figure shows, 80 per cent of these were
women.

Salaries of men and women in housing associations

Salaries for full-time and part-time employees were analysed
separately. The pattern for full-time staff is shown in Figure 9.3.

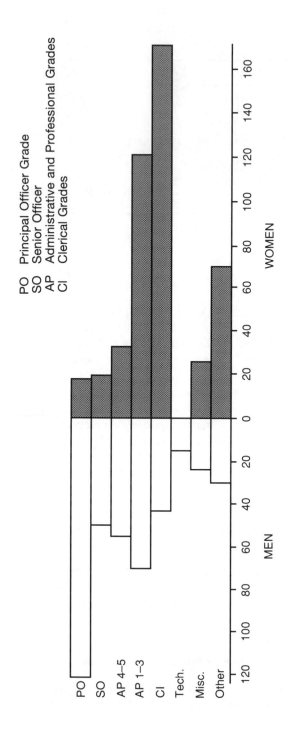

Figure 9.1 Distribution by grade of men and women in a sample of twelve housing organisations
Source: Housing Staff Survey, City University (1977) from original data

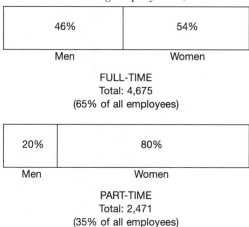

Figure 9.2   Men and women in full-time and part-time employment in housing associations, 1984
Source: NFHA survey of Housing Associations (1984)

The NFHA data, like the Staff Study, showed clearly the crossover point where a near equality of men and women employed changed to a majority of men. In housing association terms this was shown between the band £5,001 to £7,500 (52 per cent men to 48 per cent women) and £7,501 up to £10,000 (66 per cent men and 34 per cent women).

The information produced by the NFHA survey was even more striking because people working in housing associations had often assumed that women were better off there than in housing departments. Despite eight years between the surveys and different methods, the NFHA and Staff Study surveys showed similar patterns. Even those women who had managed to get through to the £15,000-plus level were in a different position from the men – they were more likely to be the sole senior employee in a small association rather than one of a number of employees at that level in a large association.

The information on full-time employees quite clearly demonstrated that women were not just simply on the lower salary bands because they worked part-time. But the differential between men and women was maintained in part-time employment: 82 per cent of the lowest salary band were women, while in the band earning £5,001 to £7,500 only 45 per cent were women. The two part-time employees earning £12,801 to £15,000 were, not surprisingly, men.

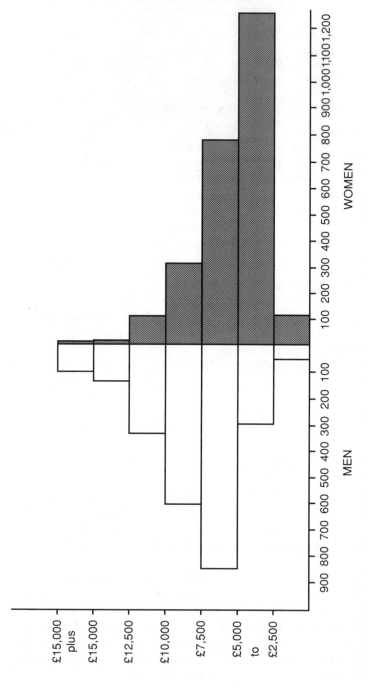

Figure 9.3 Distribution by salary band of men and women in housing associations, 1984 (full-time employees)
Source: NFHA Survey of Housing Associations (1984)

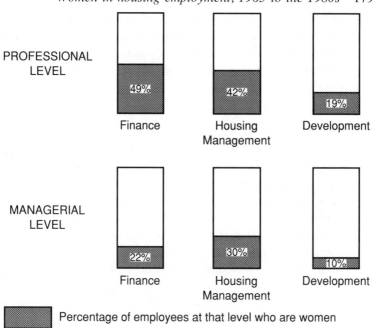

PROFESSIONAL
LEVEL

| Finance | Housing Management | Development |

MANAGERIAL
LEVEL

| Finance | Housing Management | Development |

Percentage of employees at that level who are women

Figure 9.4 Comparison of the proportion of women at professional and managerial levels in three functions of housing associations
Source: NFHA Survey of Housing Associations (1984)

The types of work done by men and women in housing associations

The survey once again showed differences between men and women not only in job level but also in the type of work done. Figure 9.4 illustrates this relationship in three selected functions of housing association. It had often been assumed that the majority of professional staff in housing associations were women, but this survey showed that by 1984 they were in a slight minority at professional level. The higher percentage at managerial level in housing management might however reflect the long history of women's involvement in housing management. Development provided the most striking figures, with only 19 per cent of the professional staff and 10 per cent of the managerial staff being women.

Hakim's national data confirms predominance of men in the building industry – women formed only 0.3 per cent of those employed there in 1971. It seems therefore that the traditional pattern of the building industry is reflected in the development sections of housing.

In the welfare function a huge bulk of women were in the 'care' category of work, a few in professional work and none in management. This reflected the traditional predominance of women in caring jobs and in jobs such as wardens of sheltered housing or assistants in care homes.

Comparison of 1976 and 1984 studies and their implications

Statistical considerations ensure that it is not useful to investigate the reasons for small differences in the findings of the two surveys. The most important feature is the consistency between the two surveys in the picture given of women's disadvantage in housing employment.

The surveys confirmed that women were generally on lower grades or salaries than men, had less responsibility and opportunity for leadership. In addition there was a tendency for women to be better represented in functions such as housing management than in development. Thus both vertical and horizontal segregation were present in housing. The women in the NFHA survey appeared to be doing slightly better than those in the Staff Study in terms of grade but it was impossible to tell whether this difference was due to sampling differences, the passage of time, or the difference between departments and associations.

These two studies provided massive evidence that the changes within the Institute of Housing had not been a fluke. They showed that few women occupied the senior jobs in housing. Taken together with the data about Chief Officers they show a long-term trend of reduced female participation at senior level. We can now see that the changes in membership of the Institute reflected changes which had been going on in employment, as well as dynamics within the Institute.

The imbalance of women's employment between different functions of housing was something which had been generally known but had not been illuminated by the Institute of Housing statistics. As we will see in the next chapter, it had important links with the situation in other related professions.

These two studies provided a basis of data from which the case for equal opportunities actions could be argued. But they also raise the question: why had the post-war period been so disadvantageous to women in housing employment?

DISADVANTAGES TO WOMEN IN HOUSING EMPLOYMENT 1945–84

Change in the size of housing organisations and local government reorganisation

It seems likely that change in the size of housing organisations was the most important single factor in the decline of women's influence. The statistics that suggest this most strongly are those which relate to Chief Officers, especially Chief Officers in London, and those which relate to housing associations.

The period of growth in the size of housing departments from the 1950s coincided with the decline in the number of women Chief Officers in London and elsewhere. Similarly, a large expansion in properties being developed by housing associations took place over this period and the proportion of Institute of Housing Fellows in Housing Associations who were women declined, as did their influence on the NFHA Council.

When local government was reorganised and the 'new' housing associations created women faced again the difficulties they had encountered in the 1930s:

> 'It was an accepted thing in X that there were women housing managers and there had been since . . . and we were given the responsibility although we weren't paid properly. But after amalgamation the other two boroughs had been mainly men and I think they thought we were a rather peculiar kind of animal – and you know they just couldn't take it that we knew as much about housing management and building construction as they did – they thought it was most odd and it took them some time to realise it . . . it wasn't hostility it was just sort of blank surprise in a way.'
>
> (Interviewee who worked through London local authority reorganisation)

Women's access to management roles

It was shown in Chapter 8 that stereotypes of women include many likely to disadvantage women as managers, for example that they are less confident, tend not to delegate and concern themselves too much over detail rather than broader issues. The career break caused particular disadvantage in management careers because 'high flying' men were getting their first steps up into management positions just at the age when many women in the past left employment to have children. Women returners were considered 'too old' for these positions despite their capabilities, as the position was based on a male model. This has been so well established that by the 1990s higher-paid women were more likely to consider paying highly for child care in order to overcome this disadvantage without sacrificing family interests.

Many studies have shown that it is often the 'old boy network' – perpetuated by drinking in the pub together or a common educational background and continual covert undermining by sexist jokes and language – which is the most powerful process by which women are excluded from the upper reaches of management (see, for example, Clark, 1991).

The kind of total dedication to work and sacrifice of all other interests expected from management staff in some organisations is a factor which still disadvantages women with dependants. More questions are being raised now as to whether this is a logical or healthy way to run an organisation for anybody. For example an article on the BBC comments:

> Married women are further disadvantaged by the unwritten assumption that those who wish to get on will be prepared to put the BBC at all times unreservedly first. The workaholic syndrome flourishes, marriages don't. What real evidence is there that the number of hours worked is related to efficiency and productivity and the best decision taking?
>
> (Meade-King, 1986)

Such voices are likely to be silenced in a recession when the emphasis is more on competitive overwork in order to demonstrate commitment to the employer. Women attending the NFHA Women's Conference felt that this was increasingly the case in the late 1980s and 1990s.

Evidence from comparative studies

One or two studies in other professions provide material which is very useful for comparison with housing. Walton's (1975) study of women in social work is especially apposite because it shows how social work had been a largely female profession to begin with but that men had been drawn more into the work as the scope widened. Higher salaries and larger organisations had attracted more men in the post-war period as organisations grew larger.

The factors Walton identified as important in this change are very close to the ones we have identified for housing. The expansion of the large housing departments after reorganisation and the 'new' housing associations had created jobs which had much higher salaries and were therefore more attractive to men. Once more men were in competition, discrimination would operate to ensure that they got appointed to more of the top posts.

Another professional field in which similar developments have been studied is that of education. Byrne (1978) showed that the proportion of principals of colleges of education who were women fell from 63 per cent in 1965 to 43 per cent in 1973; deputy principals from 68 per cent to 41 per cent. This coincided with the move to larger, urban and co-educational institutions. Woodall, Showstack, Towers and McNally (1985) examined the way in which women's promotion opportunities in universities and polytechnics were still blocked. Cooper *et al.* (1985) discuss the continuing low status of women's jobs in Adult Education from a similar viewpoint. There is sufficient evidence to suggest that the move towards larger organisations and higher salaries was a significant factor in depressing the position of women in housing in the 1970s.

Differing views of housing management and housing administration

As we have seen in Chapter 8, assumptions were made about the influence of the Octavia Hill tradition which were out of date but which may have disadvantaged women. In addition, it does seem to have been the men who were at the forefront of the pressure for comprehensive departments. This is not surprising considering that we have already established their greater concern with status. One Institute interviewee put considerable emphasis on this. He felt that Society members saw the job as essentially one of human relations between the organisation and tenants and had no particular aspirations to become Directors of Housing.

The women interviewed often agreed with the overall aims of the comprehensive housing service, but they were very conscious that the increased complexity of scope added to the problems of large size. As we have seen, the argument that this development overreached the capacity of the organisations to deliver a good service seems to be borne out by later events. But this realistic view of the limitations of housing departments would have disadvantaged women in the 1970s.

The downplaying of the 'welfare' role of housing work which was common in the 1960s and 1970s was also likely to be disadvantageous to women's participation through the operation of stereotypes.

CONCLUSIONS

There is now fairly general agreement that the causes of women's disadvantage in employment are complex and interrelated.

> Economic, psychological, social and cultural factors interact to produce the manifold patterns of discrimination that we have today.
>
> (Silverstone and Ward, 1980: 7)

> Central to much of this [research] has been a concern with the relationship between home and work and the way in which women's position in society in general, and in the labour market in particular, is influenced by and influences their reproductive role and their role in the domestic division of labour.
>
> (Martin and Roberts, 1984: 1)

Chapters 8 and 9 have demonstrated how the intergroup and individual factors which disadvantaged women within the Institute were located within a context of change in women's employment in housing.

Many different theoretical approaches can be used to illuminate these issues. Cooper (1989) distinguishes two broad traditions which have been most influential: radical feminist theory and Marxist/ socialist feminism. 'Radical feminism sees women's subordination as the fundamental form of domination in all societies . . . hierarchical power relations between men and women permeate every aspect of society.' In Marxist/socialist approaches, 'The contemporary capitalist system is seen as structured by male domination, which is organised by the capitalist division of labour' (Cooper, 1989: 25). Connell argues that it is more fruitful to consider how gender relations are currently organised to maintain the status quo:

> Instead of unanswerable questions about ultimate origins, root causes or final analysis [such studies] pose the question of how gender relations are organised as a going concern. They imply that structure is not pre-given but historically composed . . . [Such studies can also examine] different degrees to which the structuring (of inequality) is coherent or consistent, reflecting changing levels of contestation and resistance . . . Femininity

and masculinity as character structures have to be seen as historically mutable.

(Connell, 1987)

Analysing how relationships between men and women within the Institute were organised as a going concern within a historically composed situation seems to be a fruitful approach for this study. Attention can be given to identifying how processes of discrimination operate as an ongoing concern in a given occupation, without losing the perspective that these are part of the broader system of male domination in society. Inconsistencies, which certainly exist, can be taken on board rather than ignored. Concern with how oppression originally arose may be less productive than concern with how it might be changed, as long as the latter does not underestimate the size of the task to be carried out. A practice-based approach of this type provides a good framework in which to examine both the general factors affecting women's employment and those specific to housing. Connell's emphasis that the underlying issue is one of power is well substantiated in the analysis of events in housing.

This chapter has demonstrated that shifts in the distribution of power and influence within housing organisations and in their relation to the external environment had a strong influence on women's employment. Even before local government reorganisation the growth in size and scope of housing departments was increasing their power and the tendency for men to get the top jobs. Local government reorganisation and the comprehensive housing service reinforced this tendency. In addition, the stereotype of women as mainly concerned with welfare operated against them at a time when the emphasis was on quantity and cutting costs. At least some of the senior women had justifiable doubts about the extent of the move away from intensive management and about the capabilities of the new expanded housing organisations. This may have inclined them to caution about some of the new appointments. They have in many cases been proved right about these issues but at the time this was no advantage to them in the struggle for management jobs. Stereotyping and discrimination were the devices used to ensure that privileged men gained and maintained control of positions of power both within the Institute and in employment.

10 Women in associated professions

The surveys examined in Chapter 9 indicate that horizontal occupational segregation affects women within housing organisations. Women are over-represented in sections to do with welfare and care, severely under-represented in technical and development sections and under-represented at professional and senior level in finance. It seems relevant therefore to look at some of the professional bodies concerned with building and finance and assess whether women are disadvantaged in these professions and, if so, what has been done about this. This enquiry was carried out in autumn 1984 and again in spring 1992. The organisations chosen were:

The Royal Town Planning Institute (RTPI)
The Royal Institute of British Architects (RIBA)
The Royal Institution of Chartered Surveyors (RICS)
The Institute of Chartered Accountants (ICA)

All these organisations kept statistics of the numbers and proportions of women's membership (except the Institute of Housing in 1984). Research had been done on this issue in the RTPI, RIBA and ICA. The 1983–84 and 1991–92 statistics for each professional body are shown in Table 10.1. In order to make comparisons possible, corporate members were distinguished from students. The corporate member figures are examined in Table 10.1 and student figures in Table 10.2. The total number of members in 1991 is given to indicate the relative size of these organisations.

What can be concluded by comparing the figures for each professional body? Wigfall argues that occupations close to the building industry are seen as men's jobs.

Architecture, much like law, engineering and business, has always been thought of as a man's field. The architect traces his origins

Table 10.1 Women's membership of professional bodies related to housing, 1983/4 and 1991/2 (excluding student membership)

Organisation	Total members[1] 1983/4	1991/2	Percentage women 1983/4	1991/2
Institute of Housing	2,407	4,408	20.0	38
Royal Town Planning Institute	8,730	10,906	10.5	15
Royal Institute of British Architects	26,513	27,940	5.6	7[2]
Royal Institution of Chartered Surveyors	51,385	64,897	1.5	4
Institute of Chartered Accountants	79,367	97,720	6.2	12

Source: Professional organisations' statistics.
[1] Qualified members, excluding students – usually Corporate Members and Fellows.
[2] The equivalent figure for *registered* architects is 7.4% in 1984 and 10% in 1991.

Table 10.2 Professional bodies associated with housing: percentage of students who were female

	Women as percentage of registered students 1983/4	1991/2
Institute of Housing	40.0[1] (1,076)	52.0 (1,880)
Royal Town Planning Institute	24.0 (797)	35.0 (1,099)
Royal Institute of British Architects	n/a	n/a
Royal Institution of Chartered Surveyors	7.0 (1,506)	16.1 (2,007)
Institute of Chartered Accountants	n/a	35.0 (5,801)

Source: Professional organisations' statistics
[1] The 1993 figure is distorted by 'non-studying' students. In 1981–82 women formed 48–50% of students qualifying (see Chapter 11).

back to the master masons and carpenters of pre-seventeenth century days. . . . Working, as the architect did, alongside the builder, it was inconceivable that a woman should contemplate taking on such a job.

(Wigfall, 1980: 51)

So, one would expect to see the proportion of women in each occupation reflecting the closeness or otherwise to the building industry. To some extent it does. The surveyors are, after all, the group who work most directly with the building industry. But this does not seem totally to explain the difference between the surveyors and the architects, or why the planners come off so much better. The ratios of improvement are also very different. In order to examine these differences the background to the employment of women within each of these professions and the development of action for equality will be reviewed.

PLANNING

A Women and Planning Working Party was set up in November 1983 and its report produced in 1985. The Working Party paid considerable attention to issues concerning planning for women but was also concerned about women in planning employment. The poor representation of women in RTPI governing bodies was highlighted as well as the disadvantages suffered by women in employment. For example on the RTPI Council in 1982 were 51 men and 1 woman (2 per cent) (RTPI North West Branch Women and Planning Working Party, 1982: 48). 'The number of women holding office in the Institute is not representative of the size of their membership or indeed Society as a whole.' This provides an interesting comparison with the Institute of Housing Council data discussed earlier, again demonstrating how a minority (women were at that stage 11 per cent membership) comes to be even more under-represented in an elected body. Caring responsibilities and difficulties in travelling were seen as barriers to women's advancement, but factors within the professional organisation were also considered important: 'As the profession is male dominated, the structures and activities are inevitably geared towards the needs of male members.' The experience of female RTPI members in local branches can be compared with that of women members of the Institute of Housing prior to 1980. 'Incomprehension, sexual innuendo and passive resistance from some male members of the Branch Executive'; 'RTPI not frequented or supported by many women: inflexible career structure deters membership, inhibiting masculine atmosphere and rituals of Branch' (RTPI North West Branch Women and Planning Working Party, 1982: 50).

The RTPI Women and Planning Working Party of 13 members which met 1983–86 had a remit to examine

(a) the needs of women within the planned environment;
(b) the role of women within the planning profession.

An issues report on (b) was presented in 1986 in order to ensure a programme of action.

Key recommendations approved by the RTPI Council in 1987 included:

- greater flexibility in working arrangements;
- career management;
- promotion of women's interests;
- favourable membership subscription policy;
- organisation of courses for updating, career development and general Continuing Professional Development;
- favourable rates for purchase of distance learning packages;
- job sharing;
- careers information, to be rewritten.

(Howatt, 1987)

Particular attention was given to co-ordination within the Institute itself.

It was recommended that the Institute appoint representatives to be responsible for coordinating women and planning issues and activities, including the collection of information, and to report back to Council. The branches have been asked to appoint coordinators.

(Howatt, 1987)

An employment guidance note issued by the RTPI in 1988 reinforced many of the recommendations (RTPI, 1988). It quoted EOC guidance on educating employers in equal opportunities practice and advocated innovation in working practices such as job sharing and flexible work. It discussed maternity and paternity leave, temporary retirement and retraining schemes and the importance of training for women. It also discussed women's groups, sexual harassment and assertion courses. By 1992 10.2 per cent of Council members were female and

It is a matter of Institute principle that where possible at least twenty five per cent of members of each Panel should be women . . . with the ultimate aim of achieving parity between the sexes. . . . The Institute is actively seeking to encourage women into the profession and to encourage women planners to become active in the Institute. This is done through the establishment of a women's panel. . . . Equal opportunities is also reflected, where appropriate, in the Institute's literature and there is an obligation in the Institute's Code of Conduct regarding equal opportunities.

(Clemens, 1992)

In 1986 the RTPI modified the 1982 Education Guidelines and required planning schools to pay particular attention to 'the varying planning needs of specific groups, including women, ethnic minorities, the elderly and the disabled'. This was quoted in a 1988 report by Hillier, Davoudi and Healey, who observed that redressing the gender balance in planning was likely to take some time, planning schools needed to take more positive action and it was important that this was done as part of the profession's response to a multicultural and multidimensional society.

By 1992 the Institute was working with the Local Government Management Board on a survey of all local planning authorities in England and Wales, which would include details of staffing and women's employment, and had planned a similar survey of the rest of the British Isles and consultancies to follow that.

ARCHITECTURE

The first woman architect qualified in 1898, only 16 years after the qualifying examinations were instituted. Because architects have to be registered it is possible to distinguish between those registered (10 per cent were women in 1991) and those belonging to RIBA. As such statistics are not available for the other professional bodies, the tables mainly deal with members. But the fact that only 60 per cent of women architects belong to the RIBA (Perry, 1990) must be a significant argument to stir RIBA to action. The proportion of members of the RIBA who were women rose only slowly, to 4.3 per cent in 1978 (Wigfall, 1980). The increase to 5.4 per cent in 1984 and 7 per cent in 1991 continues to be quite slow (see Table 10.1). Wigfall (1980) reported on a 1973 survey of a sample of 1,015 men and women who had completed a full-time course in architecture in either 1960 or 1964 which identified a number of factors affecting women's employment in architecture. The RIBA also collects detailed statistics, for example on earning levels.

The various studies have emphasised both the difficulties which women architects have in common with other working women and those which arise from the 'male image' of architecture. In the early 1970s a number of groups began coming together in the USA to tackle the problems of women architects, but Wigfall commented that such groups had been less successful and less numerous in the UK (Wigfall, 1980: 80). However, one such group in the UK, Matrix, published a book on women's experience of architecture in 1984.

In December 1979 the position of women within architecture was raised at an RIBA Council meeting by a report from the Special Working Group of Women in Architecture. (This report itself drew on a report of the Policy Studies Institute which used statistics from RIBA.) Various steps were taken by RIBA following these reports, but in May 1982 when the position was reviewed it was clear that progress had been limited and the Women in Architecture Sub-Group was set up.

The Sub-group divided its work into four main areas, although common problems existed in all of them:

i Language
ii Careers advice
iii Architectural education
iv Practice

(Women in Architecture Sub-Group, 1984: 2)

RIBA's exposure to research and pressure is perhaps reflected in the growth in the proportion of women in the late 1970s and in the proportion of women students. For example the Bartlett School of Architecture achieved 50 per cent female entry for the first time in 1983. It was noted that in 1982–83 97 per cent of the academic staff teaching them were male, but positive steps were being taken by the Institute's Education Department to invite women to put themselves forward for inclusion on visiting boards (Women in Architecture Sub-Group, 1984: 5).

Improvement in the representation of women was, however, still quite slow and confirmed the need for a permanent group to monitor and aid progress.

> In 1985 Council accepted the report of the Women in Architecture sub group and agreed to the setting up of a permanent Women in Architecture group reporting to the Membership Committee. The ensuing five years have seen some progress, but even now only 8% of registered architects in the UK are women. However, entry of women into the schools of architecture is now 30%, compared with 18.2% five years ago, an increase which is very healthy. We need to make sure that this trend is sustained and that the profession adapts to its increasing number of women members.
>
> (Perry, 1990)

The 1990 Policy for Women Architects paper included the following recommendations:

The RIBA Education and Marketing departments should be asked to encourage young women to enter the profession using the new careers literature, the work in schools on environmental education, contacts with teachers and careers advisers and by publicising the work of women architects.

The relevant RIBA departments should remove all discriminatory language from literature, regulations, byelaws, etc. and should seek to persuade the Joint Contracts Tribunal to do likewise.

(Perry, 1990)

This paper also included recommendations for continued monitoring of progress of women students, the inclusion of women on visiting boards, free refresher courses and investigation of the changes in practice needed to accommodate women's work patterns.

Progress remains quite slow and is not currently helped by the recession. Feminists working within architecture continue to be motivated by the very direct influence which architects have on women's lives (see for example Darke, 1984) and the influence of professional mores within the built environment occupations. They have managed to set up some supportive networks and groups and seem to articulate and gain publicity for a feminist critique of the existing male domination of the profession. Improvement in women's representation still seems slower than in planning, housing or accountancy.

SURVEYING

The first women surveyors were in fact women housing managers trained by Miss Jeffery at Cumberland Market (see Chapter 5). As we have seen, for many years the RICS provided a separate examination for the Society of Housing Managers. Some women went on to become RICS-qualified surveyors and men who were RICS qualified regularly filled senior posts in the housing service. So the RICS is of particular significance for this study.

Even up to the 1960s it was on the whole 'exceptional women' who entered surveying and there was an absence of women and an absence of debate about this within the profession.

In 1967 things hot up, possibly as a result of the beginnings of the second wave of feminism and because women housing managers were having to strike back in view of their losing ground, as explained in the previous chapter. A letter entitled 'Even brighter girls' put the case for women surveyors and describes current attitudes: 'surveyors nodding and winking at the mere mention of women, and no doubt falling off ladders at the flick of a mini skirt' (Smith, 1967).

(Greed, 1991: 78)

A few more women began to enter surveying courses.

In 1973 a letter appeared in the *Chartered Surveyor* entitled 'Plumbing the depths' (Ellis, 1973) which describes a historic watershed for women. Separate lavatories for female members were introduced at the RICS headquarters. Space matters. Many women impressed on me the importance of this event.

(Greed, 1991: 79)

This could perhaps be compared with 'the affair of the Notice Boards' within the Institute of Housing (see next chapter).

In 1976, for the first time, the new president of RICS referred to women in his presidential address, asking why there were not more women in the profession 'in these days of equality' (Franklin, 1976). By 1980 'The journal actually states that the percentage of female membership had reached 1 per cent, representing a 100 per cent increase over twenty years' (Greed, 1991: 80).

There is still very pronounced horizontal segregation within the profession as well as vertical segregation (see Table 10.3). This is

Table 10.3 The Royal Institution of Chartered Surveyors: Fellows and professional associates

	1978			1984		1991	
	Total	Women		Total	Women	Total	Women
Building surveyors	2,259	13		3,396	30 0.8%	5,403	165 3.0%
General practice	10,789	241		13,079	626 5.0%	28,826	1,980 7.0%
Land agency and agriculture	3,417	11		3,812	16 0.4%	4,419	147 3.0%
Land surveyors	708	0		769	2 0.3%	1,010	19 2.0%
Minerals	504	0		488	0 0.0%	577	2 0.3%
Planning and development	951	22		1,145	33 0.3%	1,499	95 6.0%
Quantity surveyors	11,543	31		18,696	101 0.5%	23,163	393 2.0%
Total	30,171	318	0.8%	51,385	788 1.5%	64,897	2,801 4.0%

Source: Information provided by RICS

consistent with the pattern within housing, showing areas which are felt to be suitable for women and those considered less suitable. The higher percentage of women in 'General Practice' (the division which those in housing would join) remains consistent between 1984 and 1991. In the 1980s it became somewhat easier to pass from housing to surveying qualification. For example there are some degrees which qualify for both. So this may be one area where the proportion of women will continue to increase. The low proportion of women building surveyors in 1984 is probably no surprise and it is heartening to see the position improving by 1991. Greed's comment is illuminating. 'This area is often written off by women as it is perceived by them as somewhat down-market, lower class, and full of "rough beer swilling yobos"' (Greed, 1991: 92). Perhaps this is illustrative of the way in which this profession as a whole is viewed by younger women, or their parents. However, *students* in all these fields, even including minerals, are slowly increasing, as Table 10.4 shows.

The RICS did not report, either in 1984 or 1991–92, any special working parties or sub-committees concerned with the employment of women. It reported no special groups dealing with the employment of women in surveying and no research which had been done on the employment of women in surveying, though it did keep statistics. This was despite the existence of the study *Surveying sisters* by Clara Greed, published in 1991. Greed used mainly a qualitative ethnographical method based on interviews with 250 women surveyors. Her analysis also included a historical perspective on surveying and the employment of women in it as well as analysis of membership and education statistics. Was this study ignored because of its approach or do the official statements reflect the culture of the RICS?

Greed views the 'progress' of women in recent years as rather

Table 10.4 Student members, RICS 1991

	Women	Total	% Women
All students	2,007	12,464	16.1
Building surveying	173	1,975	08.8
General practice	1,086	4,338	25.0
Rural practice	56	227	24.7
Land	19	245	07.8
Minerals	11	142	07.7
Planning	58	343	16.9
Quantity	604	5,194	11.6

Source: RICS membership statistics

fragile because of the prevailing culture and distribution of power. But the women going into surveying were not necessarily aware of disadvantage or wanting to take action about it.

> It should not be assumed that the women entering the profession will necessarily hold different views from the men; nor should any automatic assumption be made that they are likely to be feminist (ironically some men surveyors assume they are and feel threatened by them without good cause), or that they will become radicalised by their experiences of being in a minority in a male subcultural group. It is a fascinating question as to why some women surveyors have become feminists, and others have not, when they appear to have had similar life experiences.
>
> Relatively speaking, women in the more private-sector-oriented aspects of the landed professions, especially surveying and to some extent architecture, are likely to be fairly 'conservative', and may appear, superficially, to have no conflict with the world of men and business. Indeed non-surveying women have commented to me that some women surveyors appear to them incredibly 'straight' and uncritical of their world view; and I have observed that those that exhibit these characteristics the most strongly are often the most successful, declaring, 'I don't think about it, I just get on with my job'. Such women are more likely to be motivated to enter the male professions by the prospect of future achievement, than in righting past wrongs or changing the built environment.
>
> (Greed, 1991: 10)

Greed sees many women in housing as having different views:

> On the other hand, women planners, and some women geographers, certain groups of women architects (such as Matrix, 1984), and many housing managers are more likely to be critical or even radical in their world view. They may campaign for the provision of better child care, especially crèches, and greater state intervention on behalf of women in society, which may actually embarrass some women surveyors, 'they're always going on about crèches, and showing us all up'.
>
> (Greed, 1991: 11)

Some of the comments Greed makes about women surveyors could equally well be applied to the more conservative housing managers.

> Many women surveyors seem quite alienated from feminism. . . .
> Yet they possess some measure of feminist consciousness of their

own, but are unlikely to express themselves in feminist jargon. . . .
They may be put off by the false media image of feminism and 'the
way feminists dress and carry on'. Since women surveyors are
operating in a minority situation, many consider it unwise to draw
attention to themselves, or to openly discuss their problems and
negative experiences, so as not to antagonise the men, possibly in
the hope of achieving more in the long run. It's a matter of 'heads
under parapets' (South Bank Polytechnic, 1987: 4).

(Greed, 1991: 11)

These observations can be linked to theories of group interaction and
strategies adopted by non-dominant groups. The 'heads under parapets' approach is typical of one of the approaches to assimilation
identified by Tajfel (1981: 332–334).

Some of Greed's comments on the isolation of women surveyors
could also be applied to women housing managers in the period from
1965 to the late 1970s. The women surveyors had a dining club, the
Lionesses, which is mentioned by Greed, but this organisation does
not seem to figure in the interviews as a source of support. Anecdotal
evidence gathered by the author would suggest that it did act as a
support for some, but was not involved in campaigning. Without
some means of women meeting together it is difficult for campaigning to start. On the other hand, Greed notes a change of attitude
among professional women in the 1990s:

There seems to be a new mood among many women in the
professions. Whereas in the past, women's problems used to be
framed in questions such as 'how can we change the profession?',
nowadays women are more likely to state, 'WE ARE the profession, let's formulate ways of reorganising ourselves and our
colleagues to enable us to operate to our full capacity as both
professionals and women.'

(Greed, 1991: 191)

Enquiries made to the RICS subsequent to the initial 1992 letter
revealed that there had been some action on equal opportunities
generally. In November 1990 the Institution's General Council
agreed to revise its equal opportunities policy, in line with recommendations made by an internal working party. A number of working
papers had been prepared for that report. They give comprehensive
coverage to the legal responsibilities and information about equal
opportunities employment practice with regard to ethnic minorities,
women and the disabled. Some of the detailed information is more

comprehensive on race than on gender, perhaps reflecting the legal situation. A small survey on recruitment and promotion policies of members' firms had also been carried out.

In 1992 *Chartered Surveyor Weekly* carried a headline 'GP division gets tough on sex discrimination'. This was in response to 'a growing number of discrimination cases being brought to light through CSW's letters page . . . also . . . revelations in the CSW/RICS/MSL salary survey that female surveyors are paid 10–15 per cent less than their male colleagues' (Morgan, 1992). The statement that 'Property has been the slowest of all UK professions to adjust its male-dominated traditions' (Goodman, 1992) was amply demonstrated by the survey and the letters. Perhaps the 'new mood' among women professionals was becoming more manifest in surveying.

ACCOUNTANCY

Initially accountancy had proved a difficult career for women to enter. Mary Harris Smith, a pioneer in the field, spent years trying to get admission to its professional body and was only finally admitted in 1919 (Silverstone, 1980: 19,20). It was not until the 1960s that the number of women accountants began to rise substantially and in 1977 research was undertaken by the Personnel Research Unit of the City University Business School which looked at some of the disadvantages women accountants had faced, including discrimination by employers (Silverstone, 1980).

By 1984, although the proportion of qualified accountants who were women was still only 6.2 per cent, the proportion among students was rising rapidly. For example among students who first registered in 1977 and were admitted to membership by the end of 1983 21.1 per cent were women. The Institution stated: 'According to the Institute's in house recruitment advisory service . . . no distinction appears to be drawn by clients requiring the services of Chartered Accountants as to whether they appoint men or women' (ICA, 1984). The improvement was demonstrated in the 1991 figures, with 12 per cent of members and 35 per cent of students women (5,414 female students and 9,911 male students).

A survey of 500 male and 500 female members was carried out in 1989 for the Recruitment Trends Study Group of the ICA with the aim of examining issues affecting the recruitment and retention of women and resulted in a comprehensive report (Silverstone, 1990). This found that in general the level of job satisfaction of both male and female accountants was high. But half the women said that responsibility for children had affected their careers, though only 22 per cent worked part-time; 12 per cent were not currently in paid employment. Most women accountants were returning to work quickly after having a child, but working part-time. More than 40 per cent of women accountants said they had experienced a number of employment problems – 41 per cent felt equal opportunities for career advancement was a problem, 29 per cent felt they had experienced sexual discrimination in training or work, 40 per cent felt employers' attitudes to women accountants needed improvement. One section, about men's and women's attitudes, is of particular interest.

> Men and women agreed on most of the professional issues. They disagreed about most of the issues concerning women. Men generally believed women with children were less committed to their jobs, that no special arrangements should be made to help them cope with a dual role, that part-time partners could not meet client needs and that the government should not provide incentives to encourage women to return to work. Both men and women believed that women have to be better than men to progress in their careers and that senior women have to choose between having a child and having a family.
>
> (Silverstone, 1990: iv)

The Institute did not entirely escape opprobrium from women members:

> 'I still have somewhere in my possession a letter from the institute, in reply to one of mine explaining that I couldn't do an exam due to my pregnancy, which started "Dear Sir".'
> 'As a major professional body the ICA should have been a leader in the field of employment of mothers but prejudices have persisted. Women's commitment is no less than their male counterparts. Their earnings are not pin money as is generally thought. In my own case I am the sole breadwinner and will continue to be. . . . Lack of support from my professional body meant that my second maternity period was unnecessarily anxious and daunting. This questionnaire suggests attitudes may be changing and that would be very welcome.'
>
> (Silverstone, 1990: 62)

The researcher commented that there had been a significant change over the ten-year period:

At an impressionistic level, the main difference between the surveys of 1978 and 1989 was the general tone of the responses. In the first survey women seemed proud and delighted to be working within the profession and conveyed a feeling of confidence and equalitarianism. In 1989 by contrast the impression gained was one of underlying anger, resentment and disappointment that attitudes had not changed, and that the atmosphere was unsympathetic, even hostile in some cases.

(Silverstone, 1990: 77)

The proposals put forward included action by the ICA to educate employers and advice on good practice for equal opportunities. An advisory service for women was popular but 'There was relatively little support for setting up a Women in Chartered Accountancy group although similar organisations in other professions have been very successful in helping to bring about change and acting as a support network' (Silverstone, 1990: 79). Subsidised retraining courses were another possibility plus greater flexibility regarding employment from both the ICA and employers. Many of these ideas were incorporated in a general report on recruitment in the 1990s published by the ICA (Silverstone, 1990).

The position of women in accountancy is of particular interest to this study for two reasons. First, the Staff Study showed that women were heavily predominant in the lower grades in Administration and Finance sections but that there were very few in the higher grades, while the NFHA study showed women in a rather better position. Secondly, and more generally I have argued that women's perceived weakness in finance was one reason for them being disadvantaged in the management of larger organisations. This indeed was the traditional view. Silverstone (1980: 24) comments: 'One respondent remarked "the name chartered accountant has always conjured up an image of a man in a pin-striped suit with a bowler hat."' Possibly this image is now becoming dimmer, but what factors contributed to the relative success of women in this field, and are the gains real and permanent?

Both the 1979 and 1989 studies indicated that some women felt that there was still discrimination which hampered their progress, and there was still a concentration of women in the lower grades. It is of interest to note that the availability of part-time work was an attraction to women at the recruitment stage (both studies). Silverstone identified two major problems which needed to be tackled – the lack of subsidised re-training courses, and the need for employers to offer promotion to women on the same grounds as men – but in the conclusions optimism was expressed about the position of women in accountancy. As we have seen, there was a greater degree of resentment among women by 1989 because employers' and colleagues' attitudes were still often prejudiced. It is possible that this goes alongside the greater confidence and greater numbers in the profession.

COMPARISON OF ACTION IN THESE PROFESSIONS

It is useful to identify what is similar in the experience of women in housing and these related occupations as well as what is different. What is it about surveying and architecture which has made progress slower than, for example, in accountancy? What are the main elements of action needed to improve equality in these professions? Given the nature of the evidence, conclusions must be tentative but the process of comparison does produce some useful insights.

All these occupations started off as more male dominated than housing and all remain so, though the gap is closing. In all of them women suffered not only from general disadvantages in employment but also from the feeling that this particular occupation was not 'fit work for women'. All the occupations which have taken action demonstrate that there may be quite a long warm-up period where issues may be raised without action being taken, or the action taken may be ineffective. It is clear that informal networks or organised groups can be of considerable help in getting equal opportunities on to the agenda and keeping it there. But such groups seem to have been more significant in housing and planning than in surveying and accountancy. Both Wigfall and Greed provide some interesting reflections on the way in which many women in private-sector orientated-parts of surveying and architecture are more conservatively inclined and do not want to 'rock the boat' by taking action for equality. The same might be expected of women in accountancy and this is somewhat confirmed by the fact that they did not wish for any organised women's group or committee to be brought in. However, it seems that in accountancy taking limited action on equality was seen as benefiting the professional organisation rather than 'rocking the boat'. It was the ICA itself which took the lead in research. Even in the early 1980s it was carrying out detailed analyses of statistics to compare male and female success rates in the examinations. This may have been connected with the fact that there was still room for expanding professional recruitment and an awareness that there would be competition for the pool of talent. The ICA (Silverstone, 1990: 21) commented on the period 1977 to 1987: 'Growth in overall Institute membership has primarily come from women's increased interest in working as chartered accountants.' On the other hand, as Table 10.1 shows, growth in overall membership of RICS from 1984 to 1992 was still male-led.

Greed considered that women in housing were more likely to take collective action because housing had a more socially conscious tradition. But this had mixed effects. Women housing workers with more radical views often did not want to take the professional qualification or to join the professional body. The more patriarchal the Institute of Housing appeared the more powerful this factor was. Also there were many women with more conservative views within the Institute, and feminist groups had to take account of this. On the other hand those feminists who felt it was necessary to deal with power structures were willing to work within the Institute and once the Institute began to appear more responsive to equal opportunities issues this encouraged more women to join. This issue is more fully discussed in the next chapter.

Although the women surveyors had a group this does not seem, in the period under discussion, to have taken on a campaigning role. Greed comments on how isolated women surveyors are, so they may need this kind of support. It is possibly a combination of the 'not rocking the boat' ethos and the sheer size of the profession that makes such organisation difficult. The incentive of increasing the professional membership, which seems to have operated at different levels with the ICA and the Institute of Housing, does not appear to have been so powerful with the RICS. Did it feel that it could still recruit enough men?

The initial recruitment of women to an occupation is affected by its public image. Is the builder's navvy a more powerful male image than the pinstriped City gent – especially now that many professional women adapt the City gent's uniform quite successfully? Surveyors are associated with being out on building sites and getting muddy (even though the RICS itself has to some extent been fighting the tin hat image). The architect's popular image is of someone sitting at a drawing board, and planners are popularly seen as even more remote from mud and toil. Perhaps it is easier for a woman to adopt pinstripes and still be seen as 'feminine' than it is to adopt the muddy boots. Women in the manual trades have certainly found that this kind of image is regarded as unfeminine (see, for example, Women and Manual Trades, 1988). Many teenage girls are still heavily influenced by stereotyping and peer pressure towards keeping clean and tidy and 'feminine' in appearance.

Both in housing and surveying some women were deterred from taking equal opportunities action by a stereotyped view of feminists. One of the roles which organised women's groups can play within professions is to demonstrate that feminists can be as diverse in dress, style and behaviour as most human beings are.

Class and power

Class and power may also have an influence. Architecture has a strong traditional upper/middle-class image – this may be influential with parents, who view it as a suitable occupation for their girls. The somewhat less classy, 'rougher' image of surveying might be seen as less suitable. Feminists have drawn attention to the effect of architecture and planning on the environment and this motivates some women to overcome the difficulties of such a career. The influence of surveying, though also powerful, is rather less evident to the general public.

The persistence of vertical occupational segregation confirms that barriers to women increase with increasing amounts of power. Many writers are beginning to emphasise that power is a key issue (for example, Stacey and Price, 1980b; Eisenstein, 1984: 139; Coote and Campbell, 1982: 241–248; Rendel, 1981). But this explanation seems inconsistent with the relationship of money and power and the increase of women in accountancy. But the gains of women in accountancy may still reflect the power structure. Women may be concentrated in aspects of financial work which call for the stereotyped virtues of neatness and attention to detail (like audit) rather than in situations where financial power is exercised.

Educational influences

The educational qualifications needed for entry to the profession have sometimes been considered a barrier. Maths has been seen as a 'male' subject, but girls have been improving their performance in maths (Smith, 1986 and EOC, 1985). Maths is needed for accountancy as well as for surveying and architecture. The subject of technical drawing, however, has a specific relationship to surveying and architecture. The acquisition of skill in technical drawing may cause young people to consider this type of occupation, or employers providing day release may look for it. Technical drawing and allied areas still have a heavy male stereotype in schools and colleges (EOC and Somerset Council, 1982). Schools with good equal opportunities policies were in the 1970s and 1980s ensuring that girls had an adequate 'taster' of technical drawing – it is obviously important for this to continue. The progress previously made may be undermined by the pressures of the national curriculum so this issue needs careful monitoring.

The way in which women are treated as students or would-be students in these occupations seems likely to be of considerable significance (see for example Women in Architecture Sub-group, 1984: 4, 5; Silverstone, 1980: 24–28). Greed (1991) has emphasised the mixed views of women surveyors about their education. Feelings that they enjoyed the course and did not want to seem critical mixed with some quite striking accounts of discrimination. The experience of female housing students in technical subjects and of girls in lower-level technical courses (see for example Leevers, 1986: 10) suggests that discrimination both by other students and by lecturers can be influential in success or failure. Wigfall (1980: 69) suggested that women's experience of being in a sexual minority was also important in architectural education. The recruitment of more female teaching staff as well as equal opportunities guidelines has been seen as important in most of these studies. Additional measures and effective ways of monitoring equal opportunities in independent tertiary institutions are discussed in more detail in Brion (forthcoming).

The recruitment of first-year students is one of the most sensitive indicators of women's ease of entry to a profession. A number of these professional bodies were, quite rightly, ensuring that this statistic was available and was monitored. Given different qualifying systems in different professions, it is sometimes difficult to get directly comparable statistics. Table 10.5 shows an interesting and on the whole encouraging picture, but RICS statistics are not available.

Table 10.5 Professional organisations related to housing: women as a percentage of new entrants, 1991/2

Royal Town Planning Institute	40.3
Royal Institute of British Architects	30.3
Institute of Chartered Accountants	34.5

Source: Professional organisations' statistics

CONCLUSIONS

There is a fascinating range of differences between these professions in respect to women's employment. It is clear that the easiest way to get action is to convince a professional body that it is in its interests to ensure equality of opportunity. But this argument can be put to any profession: why are there differences in the rate of change?

It seems likely that the most significant differences between surveying/architecture and the rest include the image of the building industry, the perception that women face higher risks of discrimination, sexual harassment and barriers to promotion within these professions and barriers to women's access to technical drawing at school level. By the 1990s the effects of the recession on the property-related professions were a further influence. There has been more action on equality of opportunity in architecture than in surveying but the seven-year architect's training with little chance of earning extra income is probably the significant comparative factor continuing to deter women from studying architecture.

It would not be easy to establish whether the incidence of sexual harassment and discrimination is in reality greater in the landed professions than in accountancy or law. It is more overt, especially in the building industry. But in every occupation which has taken action for equal opportunity plenty of evidence about harassment and discrimination has come to light. Some female lawyers might argue, for example, that discrimination is more covert and subtle in their occupation, but as common. Recent cases regarding the police and medical practice have demonstrated that professional women may find many hidden barriers. But it also seems possible that a vicious circle could develop. Because the statistics now demonstrate that the landed professions have made slower progress to equality, female entrants may think twice before joining them.

Given the sophistication of research now available and media knowledge about image-changing it can be argued that the means are now available for the landed professions to take more effective action in encouraging female membership. In all these professions, including housing, the promotion of women already in employment and their representation within the professional organisation remain key targets which must be tackled during the 1990s. Some issues arising from relationship with land and building are common to housing, architecture, surveying and planning. In recent years there has been more activity to co-ordinate equal opportunities in the built environment professions through such groups as WICAG

(Women into Construction Action Group), WEB (Women and Built Environment) and the Women's Design Service. Public funding for such issues has been diminishing but it is important to women that some support should continue.

This review of action by some professional organisations together with the history of women in housing employment makes it possible to provide a checklist of action which a professional organisation can undertake. This is arranged under a series of questions related to career stages.

A checklist for action by professional organisations

1 Do women apply for entry to the profession or courses?

Statistics

— Applications and acceptances at college and student membership department level.

Action

— Formal statement of equal opportunities by the professional organisation, keeping, reporting and evaluation of monitoring statistics.
— Action in the educational system and the media on women's aspirations in general and removal of restrictive stereotypes.
— Specific action on recruitment literature and examination of channels of entry, especially where day release or lower-level courses are involved.
— Ensuring that there are female teaching staff.
— Ensuring that women are involved in presenting information and advising students.
— Use of non-discriminatory language and behaviour in the educational process and in the profession.
— If there are substantial problems at this stage positive action may be required, e.g. women only access courses, articles in the women's press, talks at schools, etc. Such action needs to continue over a period of time, with effective feed in to the main career streams.

2 Do women qualify?

Statistics

— Entry and pass rates. Feedback from students and newly qualified members.

Action

— In the post-war period most professions have found that women's pass rate is similar to that of men. But it is also important to identify whether discrimination in the educational system deters women from continuing in the profession, depresses their aspirations or channels them into particular types of work. Implementation of equal opportunities in the content and process of teaching can help to change the culture. The presence of supportive female staff can be an important source of help, especially if women are in a minority.

3 Do women get a fair allocation of first jobs?

Statistics

— Employment rates of qualified staff. Identification of any horizontal discrimination within the profession.

Action

— Implementation of equal opportunities employment practice. Guidance on this from the professional organisation. Comparison of statistics from different employees/regions (see NEDO and RIPA, 1990).

4 Do women stay in the occupation?

Statistics

— Membership statistics, cohort studies.

Action

— Employment conditions, maternity, paternity and dependency leave, job sharing, return to work, etc. (See Chapter 11).

— Tackling problems of subscription rates to the low paid, professional insurance issues.

5 Do women get promotion?

Statistics

— Employment studies comparison with other professions. Professional organisation statistics may help, for example, if there is a Fellow grade or equivalent.

Action

— An absolutely crucial issue which needs long-term and sustained action.
— Education of employers and the profession itself. Support for women if legal action needed (see, for example, Hansard Society, 1990). The professional organisation can have influence by setting an example (see item 6), initiating research and monitoring studies, discussing and advising on action.

6 Do women participate in the professional organisation?

Statistics

— Women officers, committee and branch members, women staff and their grades.

Action

— Examination of the mechanics of election, recruitment and promotion of committee members and staff. Target setting.
— Guidance on non-discriminatory language and behaviour.
— Review of location, timing and conduct of meetings.
— Ensuring that conferences and training are 'women friendly'.
— Targets for proportion of female speakers at public events.
— Ensuring that all policies are examined for equal opportunities implications.
— Targets for proportion of women on consultation panels, visiting and validation boards etc.

It seems that in each occupation the case for equal opportunities has to be fought through separately and the statistics demonstrated – and this may need to be done at five-year intervals. One of the hardest messages to get over is that doing nothing does not constitute fair treatment if one group is currently in the minority; it merely reinforces the status quo. Guidance given must be specific not only to the occupation but also to particular contexts. For example, in a professional group known to the writer men dominated discussions. Although many of them were 'experts' in communication they assumed that equality existed because there was no conscious intention to discriminate. Evidence from Spender and others about male/female communication patterns was presented and group members challenged to check it out within their own experience. A number did so and became much more aware of how male dominance in language worked and how behaviour had to be changed within the group to ensure an equal hearing for the women.

The role of professional institutions

Many of the measures discussed are directly under the professional organisation's control, for example orientating recruitment and marketing of the profession to include women. Recruitment is also affected by employers' policies for day release and practice in the colleges. Most professions now have a process of accrediting educational organisations for professional qualification so it is possible to include monitoring for equal opportunities there, as BTEC and NCVQ do.

Employment and promotion patterns are less directly controlled by professional organisations but many of the organisations studied had issued guidance to employers on this (e.g. Institute of Housing, RTPI). It is often useful to explore employment issues in conjunction with other organisations concerned – the Planners were working with the Local Government Management Board; the National Federation of Housing Associations was a major channel for recommendations to housing employers. Return to work or support for those temporarily retired could also be helped by the professional organisation or organised on a collective basis (see Chapter 11). The profession's Code of Ethics can also be strengthened, or implementation emphasised.

In this context the relative lack of action by the RICS becomes more glaring. It remains to be seen whether Opportunity 2000 will have more impact than previous initiatives. Though a professional organisation may not consider itself responsible for general factors in society which disadvantage women, it is certainly responsible for action within its own remit. There are no *a priori* reasons why the good practice of the professional organisations who have increased women's representation cannot be adopted by all of them.

11 Postscript: Women in housing begin to organise again

THE EMERGENCE OF NEW GROUPS

In 1965 some women felt unhappy about their position within the new organisation, but were not able to take effective action. By the early 1970s, women's presence within the Institute was at a very low level. However, there were a few former members of the Society who were in positions of influence. For example, in 1978 Mary Smith (one of the housing managers who worked for the Ministry of Supply during the war) became one of the housing management advisers at the Department of the Environment, thus reviving an old Society tradition (Smith, 1979). A few women of a slightly younger generation, who had trained under the Society, were beginning to play a more active part within the Institute (some having returned from career breaks) and finding common interests with women who had been trained within the Institute but objected to its male domination. There were other women with feminist views working in, or having an interest in, housing, often involved in the newer types of work such as housing aid, homelessness, co-operatives. Many of these women did not belong to the Institute and some were hostile to it. Outside the world of housing, feminism was growing stronger and equal opportunities policies were becoming more of an issue (Coote and Campbell, 1982).

In the 1970s women from these different traditions began to form alliances and to take action on issues affecting both housing employment and the delivery of housing services. This chapter traces the formation of the first initiatives in some detail and then briefly examines two issues as examples – return to work and education. The strategies and tactics used by the groups working with the IOH and NFHA are then evaluated.

The first moves were, not surprisingly, fairly hesitant, as women felt themselves to be in quite a weak position. On employment issues the professional organisation once again provided both the means and the focus for action. In November 1973 a letter from Mary Smith, published in *Housing*, drew attention to the lack of women on two DOE Advisory Committees and to gains and losses in women's representation in housing. On being contacted by the author in 1974 she wrote:

> I think some statistics on the position of women in housing could be very interesting. . . . I agree that some positive action should be taken about the male domination of the Institute, and of the housing field generally, and steps as suggested above could start the ball rolling.

Mary Smith later commented that the lack of women's representation on the Institute of Housing Council had caused some of the ex-Society members to 'get excited' both about this and about what seemed like almost a deliberate blotting out of the history of the Society of Housing Managers (Smith, 1987).

The affair of the noticeboards

The issue on which action by older and younger women members began to coalesce was, perhaps, symbolic. It had been discovered that no care had been taken of the historical records of the Society and, in the early 1970s, senior staff at the Institute were unaware that irreplaceable minute books of the Society were stored haphazardly in the basement of the Institute's offices at Victoria House. In addition, at that time, noticeboards erected in the main meeting room commemorated the presidents of the Institute of Housing from 1931 to 1965, and presidents of the joint Institute from 1965 to present day, but made no mention of the Society of Housing Managers. Some women members felt that this was invidious in view of the contribution of the Society to housing and to the formation of what was supposed to be a joint organisation in 1965. Strong pressure from women members was required before a new notice-board was erected bearing the names of presidents of the Society from 1931 to 1965, but contacts between the women had been renewed and reinforced.

Outside of the Institute a Women and Housing Group, which included a much broader range of women interested in housing issues, had been in existence since the late 1970s and held a conference in March 1980 at which issues concerning women as consumers and employment issues were discussed (Women and Housing Group, 1980a). It went on to run an evening class in 1980–81 under the auspices of The London University Extra-Mural programme. By the end of that year, however, it decided that its objectives and membership were too diffuse to obtain adequate support and therefore decided to wind up its existence. Later, other women and housing groups began to emerge (Metters, 1981).

THE FORMATION AND WORK OF THE WOMEN IN HOUSING GROUP

Publication of the book *Women in housing: access and influence* (Brion and Tinker, 1980) provided the impetus for drawing together women working in housing. A meeting was held in May 1981 which inaugurated the Women in Housing Group. Membership of this group was open to all women working in housing, broadly defined. It included academics and a wide selection of women in housing-related jobs as well as Institute members. The group interested itself in both employment issues and consumer issues. It gradually formed a national network with a newsletter and some local groups, the strongest of which were in London and Sheffield. By 1985, the Sheffield group had held a major conference, largely on consumer issues (Women in Housing Group, Sheffield, 1984), while the London group had inaugurated and helped to run an agency for part-time and temporary work in housing (Housing Employment Register and Advice, described later).

By the latter half of the 1980s the work of the Women in Housing Group had contributed to the formation of other groups, notably the Institute of Housing and the National Federation of Housing Associations' Women in Housing working parties and HERA (Housing Employment Register and Advice). From December 1987 the group decided that these official organisations were likely to be the main focus for action over the next few years and since all concerned were short of time it was best to keep only a skeleton organisation to ensure that a network was available when needed.

THE INSTITUTE OF HOUSING WOMEN IN HOUSING WORKING PARTY

In the London Women in Housing Group, representation of Institute of Housing members was fairly strong and it was decided to make the Institute one focus of action. By 1982 members of the Women in

Housing Group were addressing branch meetings of the Institute and putting forward a manifesto, entitled 'Women and housing. Action which the Institute of Housing needs to take' (Brion, 1982).

By November 1982 the Professional Practice Committee agreed to the setting up of an Institute Women in Housing Working Group (Smith, 1982). The first meeting was held in February 1983 (Institute of Housing Women in Housing Working Group, 1983a, 1983b). But the status of this working group was not very clear and its membership not very wide or representative.

A breakfast meeting was held at the Institute of Housing 1983 Conference and by the time Conference 1984 arrived it was possible to put forward a much more coherent programme and a request for a much broader-based working party (Brion, 1984). An official working party was set up in September 1984, composed of women from different branches of the Institute and some Council members. The early days of this working party were occupied with a continuation of the efforts to improve the Institute's response to women members and in particular with the issue of members seeking to return to work after a career break. Three publications giving advice to women members and branches eventually summarised this work (Institute of Housing, Women in Housing Working Party, 1984b, 1984c, 1984d). The working party felt, however, that these measures would not be sufficient in themselves and a funded Return to Work project described later was set up.

A code of non-sexist language was drawn up and eventually recommended to staff. The Institute's equal opportunities policy was discussed. The need for statistical monitoring of the numbers and participation of women members was repeatedly discussed but this was not finally implemented until 1987.

By 1985 the need for some updated information on women in housing employment was becoming evident; the NFHA survey, carried out in 1984–85, had naturally only covered housing associations (NFHA, 1985). A major survey of women's employment in local authorities was begun in 1985. Its findings were published as *The key to equality* (Levison and Atkins, 1987). This survey showed that the position of women in housing employment was little better than earlier surveys had found. It therefore strengthened the working party's case for remaining in existence rather than dissolving after a limited period, as was usual with Institute working parties. The new programme of work for 1987 and 1988, besides continuance of the earlier activities such as return to work and job sharing, gave more stress to issues affecting women as consumers of housing. Women and homelessness was highlighted for the 1987 conference; domestic violence and security on estates for the 1988 conference. The working party was concerned at various times with issues of relationship breakdown and housing, the repairs service, customer care, women, poverty and housing. Another major item of work in 1987–89 was originated by women members but in fact affected all working in housing; this was a guide, and eventually a survey of local authority and housing association practice, on violence at work (Poole and Porter, 1988).

Following on from the 1985 investigation into women's position in related professional bodies, described in Chapter 10, funding was sought from the EOC for a joint project with RTPI, RIBA and RICS, but this was not successful.

By 1987–88 the membership of the working party had been broadened to include women from pressure groups but it still had only temporary status. It was able to run workshops, social sessions and an exhibition stand at the Institute Conference, but the process of making the Institute Conference more 'woman friendly' was a long one and continued well into the 1990s. Publication of membership statistics analysed by gender confirmed that women's participation in the Institute was improving (Figure 11.1).

By 1989 internal issues were beginning to loom large. Most members of the working party felt that what they could achieve as a temporary working party, not fully integrated within the IOH structure and reporting only to one committee, was becoming too limited. They felt that the group needed to be put on a more permanent basis and it needed to report to Policy and Resources Committee. Strong hints were given to group members that if they asked for too much the group would be dissolved. However, problems arising from the lack of integration of equal opportunities within the Institute's structure were becoming too apparent and the group continued to raise these questions. In November 1989 'it was agreed . . . to suggest to Council that all major policy decisions should be monitored for equal opportunity implications' (Institute of Housing Women in Housing Working Party, 1989).

A period of negotiation and lobbying began and the eventual upshot was the formation of an Equal Opportunities Working Group with a clear two-year brief and a much more defined place within the Institute's structure (which was itself revised in 1991). The strategy of the new group was to concentrate on issues within the Institute in the first year and outside in the second year. One of the first pieces of work was the approval of guidelines for speakers at all

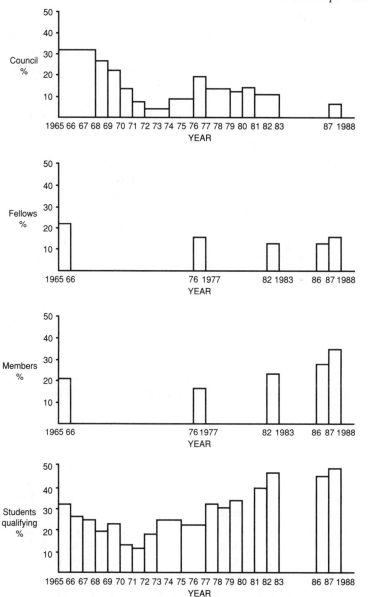

Figure 11.1 Comparison of changes in women's participation at Council, Fellow, member and student level of the Institute of Housing, 1965–88 (per cent)
Source: Data collected by author from Institute of Housing reports, yearbooks and examination results

Institute of Housing events. A mission statement on equal opportunities was adopted by 1991. By the end of 1992 professional practice guidance on equal opportunities was being drafted, the Institute's own employment practices had been thoroughly reviewed and there had also been extensive work on the implementation of equal opportunities in the Institute's education and training programme. Arrangements for monitoring and reviewing equal opportunities implementation had been much improved (Institute of Housing Equal Opportunities Working Group, 1992).

The group recommended to Council that the Institute should ensure that public representatives of the Council were chosen to give better representation of the different regions, women and other under-represented groups. Electoral arrangements for Council had proved a thornier topic, and was held over for further discussion. The adoption of targets for representation of women and ethnic minorities at Council and branch level was proposed. Action on other issues which might limit women's participation on Council was suggested. In November 1992 Council adopted a target of at least 33 per cent of Council members being women by 1995 and 50 per cent by the year 2000, and accepted the majority of the working group's recommendations.

THE NATIONAL FEDERATION OF HOUSING ASSOCIATIONS' WOMEN IN HOUSING WORKING PARTY

From 1980, the Women in Housing Group contained members in employment with housing associations as well as in departments, and it was clear that it would be beneficial if the NFHA were taking more positive action. Links were made with other women working in housing associations, and in particular with the NALGO Housing Association's Branch, which had some feminist members. Despite this, progress seemed slow until 1983. At this stage, a joint letter from the Women in Housing Group and the Housing Association's Branch (Hargreaves, 1983) coincided with other factors favourable to the establishment of a working party. A number of different discussions and initiatives were going on in the Federation at the time, including one on fair employment. A member of National Council (June McKerrow) was interested in the issue and pushed for the setting up of a working party (Stanford, 1989). Agreement was reached within the NFHA on the setting up of a Women in Housing Working Party with a two-year lifespan. It was decided that the first year of operation would be concerned with women in housing employment, the second year with women as consumers of housing.

The NFHA Women in Housing Working Party was a large one, with systematic representation from all parts of the country. It was set up as a temporary working party with limited duration. It was able with NFHA resources to carry out the major survey on women in housing employment which is described in Chapter 9. The NFHA agreed to publish the report but not to make it an official recommendation in order to avoid the delay which might be caused by need for detailed committee approval. This meant it was vital that the working party should persist in looking at implementation issues, and at the position of women within the NFHA. The group gained early agreement to setting a standard for 50 per cent female speakers at NFHA conferences; the logical corollary of this was greater training provision within NFHA for women, both on public speaking and on assertion. Encouragement was consistently given to the setting up of local groups which could both support women members and put pressure on employers. Where appropriate, these were encouraged to link with Women in Housing groups or with local Institute branches that were taking action; though housing association members were sometimes reluctant to associate themselves too closely with the Institute of Housing.

In its second year the working party went on to consider issues of significance for women as consumers of housing, producing a series of articles for *Voluntary Housing* and continuing to publicise the work through meetings and the NFHA conferences. A standing group on Women and Housing was eventually established. By 1988 they were able to set in motion the planning of a women-only conference to be held in spring 1989 (NFHA Women in Housing Standing Group, 1988).

This was an important development. Doubt had been cast on whether a women-only conference would be viable. In the event it was well attended. It was, however, the scene of a stormy confrontation between 'the platform' and black, lesbian and working-class women who felt that they were being excluded by the elitist stance of some speakers. Resolutions from the floor stated, for example, that 'There has been a lack of understanding around race issues and we feel that white women have been seen in this conference to be unwilling to accept their own racism.' 'This is a *women's* conference – it should not be divisive, but rather supportive. Minority groups should not be marginalised, nor should they segregate themselves – we should be joining together and providing support for each other, still recognising our different backgrounds' (NFHA Women's Conference, 1989). This controversy is discussed in more detail later in the chapter. Speedy efforts were made to open up dialogue with these groups and to ensure that future conferences and the working party would be more representative and responsive to their needs. Though harassing for many who took part, this confrontation proved healthy in the long run by ensuring that hitherto excluded groups became more visible. It did not, as some feared, mean an end to such conferences. In the short term, some people were put off by the thought of such open controversy but as it became one of the legends of the movement it reinforced awareness of how strongly some women felt about the multiplicity of discrimination they faced.

In 1993 the fifth NFHA women's conference was held. This conference was of considerable significance, with a wide range of issues related to women and housing being debated every year. It was attended by a much larger proportion of junior and front-line staff than any other NFHA conference, though tenants and older women were still under-represented because of financial constraints. The women's conference provided a way for grass-roots issues to be fed through to the Standing Group. The group reported to the NFHA Equal Opportunities Sub-Committee so, with other standing groups, could make suggestions for changes in policy and procedure, comment on consultation documents, etc. Only very limited funds were available for research. 1992–93 projects included single women's access to housing and a domestic violence report.

One important activity of the group was organising 'fringe meetings' at other NFHA conferences. This was a way of improving communication and sometimes observing at first hand reasons why at certain conferences (for example, Chief Executive's and Maintenance) women's attendance was low. By 1992 it was agreed that NFHA conference attendance was to be analysed and a 'women's access to NFHA' guide for planning groups was to be produced. One of the items on the agenda for this period was improved co-operation with the Institute of Housing on women's issues – it had been argued by women attending the 1992 conference that scarce resources were best used by co-operation between the two. The NFHA had also carried out a further equality survey (NFHA, 1992). As this was not designed to be comparable with the 1985 survey, its usefulness for women's issues was limited. It did however confirm that women were still in a minority in senior jobs, and in 1992 the Housing Corporation convened a 'Women at the Top' seminar and report, drawing attention again to this issue and making practical recommendations (Office for Public Management, 1992).

THE RETURN TO WORK ISSUE

The retraining schemes

The return to work issue was a good example of the way in which a broad and vital issue for many women could form the focus for successful action by the groups mentioned here, and for suitable alliances with organisations outside the housing world.

When the first Women in Housing Group meeting was held in 1981 the issue of women qualified and experienced in housing work who were having difficulty in returning to employment after a career break was raised (Women in Housing Group, London 1981a; Dallas, 1984). This was a good campaigning issue, both because of its grass-roots support and because of the economic arguments that it represented a waste of the public money and employers' money spent on training these women.

Because the Institute of Housing was the major examining body for housing, it was also a natural issue for the Institute's working party and one which would meet with response for its members. In fact most of the first two years of the working party were taken up with considering this issue. This resulted in a scheme for encouraging employing organisations to offer facilities to women wishing to return, similar to those being employed in banking and elsewhere (see, for example, Institute of Housing, Professional Practice and Publications Committee, 1984b). The Institute could only provide small resources to administer this scheme, so it was only partially successful. It was clear that more resources were needed from elsewhere.

A proposal for running Return to Work courses was put to the Local Government Training Board in 1984 and negotiations eventually resulted in an MSC-funded Return to Work Project running from May 1985. The work of the project centred around the running of two sequences of retraining courses but it also generated much valuable publicity for the whole issue.

The formation of HERA

Many of the women returning to work wanted part-time employment. There was no central employment help for housing, such as RICS and

RTPI provided for their members. The Women in Housing Group tried initially to operate a register on a voluntary basis for the London area but this proved too onerous and the legal complications were daunting.

It proved possible to negotiate with the Over Forty Housing Association (an organisation founded in the 1930s, later called Housing for Women) the setting up of a new body called HERA (Housing Employment Register and Advice). The service extended to all women working in housing, stressing that wardens, secretaries and caretakers were equally housing staff. In addition, it crossed the boundaries between housing association and housing department work, focusing on women's employment in housing, wherever it was. By 1988, HERA was carrying out a wide range of activities including jobs register, career counselling, training, information service on housing education and training, and producing its own publications (HERA, 1988).

Looking at the return to work issue overall, much of interest emerges. It was a good issue for initial action because it so easily made sense to the professional bodies and funding agencies such as MSC. A broad coalition of interests could be united in pursuing it and could each play their part. It provided experience of successful action. Finally, it was particularly pleasing that an initiative from the earlier wave of feminism in the inter-war period could be helpful in aiding development in the 1980s.

Education

On the retraining issue it was possible to work in a fairly straightforward way within the Institute of Housing. To maximise opportunities for part-time and temporary work, direct action was taken by women across all the organisational boundaries. To have an impact on education for housing work generally required much more complex, long-term and difficult processes. The Institute of Housing was in charge of the professional qualification but, over most of the period described, the Women's Working Group did not have representation on or formal input to the committee concerned with education. Lower levels of qualification were the concern of BTEC and NCVQ. Both of these fortunately had equal opportunities policies, but implementation in a particular occupation would be influenced by a variety of groups, especially employers' representatives. Only NCVQ had a specific working party on the implementation of equal opportunities in their new qualification. In these circumstances progressing equality issues in housing education relied either on members of the women's networks who were on the relevant committees or, more frequently, on influencing committee members or officers. It often involved detailed and time-consuming work assessing proposals from an equal opportunities point of view and summarising the issues as well as hard work to lobby relevant people to ensure a hearing. Some of this activity is described in more detail in Brion (1994).

The setting up of the improved equal opportunities structure in the Institute of Housing in 1990 brought about a more fundamental review of educational strategy and equal opportunities. This high-lighted areas for further work, for example increasing the attention given to equal opportunities during the validation process and ways of inputting data on equal opportunities into the IOH decision-making process (Institute of Housing, 1991).

The outcome of all these levels of work, fortunately, was a system of professional education which maintained access for people in employment and without educational qualifications, and drew atten-tion to equal opportunities in various aspects of the curriculum. Even by 1992, however, much remained to be done.

Other groups

The 1980s saw the development of women's initiatives in a number of organisations across the housing scene. Significant studies were done, rarely receiving public funding and relying on the hard work of pressure groups or individual researchers. The number is too large to cover here in detail but a few examples will indicate the range. Shelter, for example, consistently wrote women's issues into the agenda (McKechnie, 1990). Publications dealing with women's experience of housing and the effects of housing policy on women became more numerous. Groups who suffered from additional dis-advantage, such as women suffering from domestic violence, black women, disabled women, lesbians and older women, pointed out strongly that not all women were equally disadvantaged. Many of these groups wanted to play a more significant part in policy deci-sions and the delivery of services as well as getting housing provision adapted to their needs (Binney *et al.*, 1981; Welch, 1986; Clark *et al.*, 1987; CHAR, 1988; Morris, 1988; Women and Housing Group, 1984; Blair, 1993; Sexty, 1990). Articles with a more theoretical orientation were published (for example Munro and Smith, 1989; Brailey, 1987).

Table 11.1 Comparison of women's membership of selected branches of the Institute (percentages)

Branch	1987 Fellow %	Member %	Student %
London	31	35	56
East Anglia	6	24	43
South East	8	26	51
North Western	0	29	44
Scotland	12	35	53

Source: Institute of Housing statistics

Geographical spread

Initiatives were also taken in other parts of the UK – for example in Scotland, where women had been very disadvantaged in the past. The Northern Ireland Housing Executive in May 1991 published the first report of its Equality Working Group which demonstrated the existence of vertical and horizontal occupational segregation in the Executive and proposed a comprehensive range of measures to improve the situation.

In 1987 the Institute of Housing published an analysis of membership by branch and gender (Table 11.1). This is no longer produced. It was, however, quite useful since it was difficult for members in the North-West, for example, to argue that women were less capable of housing work there than in London so it undermined the case of some branches that had said 'we have no need of an equal opportunities officer'. Identifying members for less than three years confirmed how

Table 11.2 Women's membership of selected branches of the Institute, April 1987 (members for less than and more than 3 years)

Branch	% members less than 3 yrs who were women	% members more than 3 yrs who were women
London	51	26
East Anglia	30	21
South East	41	17
North Western	39	22
Scotland	49	24

Source: Institute of Housing statistics

recent some of the increases in women's membership were (Table 11.2).

These figures confirmed the substantial progress that had been made in Scotland – one of the two branches with more than 50 per cent female students.

WOMEN IN HOUSING BY 1993

1987 was the first year that the Institute of Housing produced statistics of membership analysed by gender. This was a change which had been requested by the first meeting of the Women in Housing Working Party in 1983. The detail and presentation of these statistics have varied over the years. The 1993 membership statistics confirm the trends shown in the previous chapters (Table 11.3). Female participation continued to increase at all levels in the 1990s, but in the context of growing uncertainty about housing as a career.

Promotion to senior posts remains an issue. The *Getting more women to the top* report (Office for Public Management, 1992) pointed out that even in those associations which returned to the 1990 NFHA survey and had equal opportunities policies 82 per cent monitored equal opportunities in employment but only 17 per cent monitored promotions. Practical suggestions were made to increase the proportion of women in senior positions.

It is not always necessary to rely on expensively gathered survey data. The NFHA training department collected monitoring data and this was analysed by the author in 1989 and 1992 (Figures 11.2 and 11.3). This confirms that there are still problems about women's participation at the higher levels of organisations and in the technical sections like maintenance. Evidence from women attending such conferences had indicated that they were far from 'woman friendly' and this issue was taken up by the NFHA Women's Standing Group as part of their programme. The Women in Housing Group also requested that the NFHA equal opportunities sub-committee regularly examine the monitoring statistics in future.

Table 11.3 Institute of Housing 1993: percentage of each grade who were women

Fellows	19
Members	43
Student members	55

Source: Institute of Housing statistics

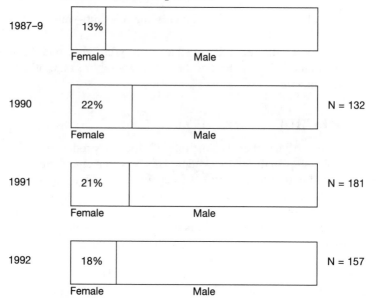

Figure 11.2   Attendance at NFHA chief executives conference, 1989–92
Source: NFHA Training Department

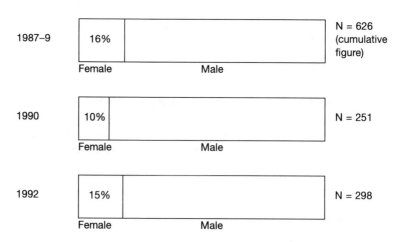

Figure 11.3   Attendance at NFHA maintenance conference, 1989–91
Source: NFHA Training Department

CONCLUSIONS

The achievements of the period

The 1980s had seen some substantial achievements, especially within the Institute of Housing. In 1980 it had hardly been considered legitimate to discuss women's position within the Institute. By 1992 open discussion could take place at all levels and had resulted in effective action. The statistics indicate that housing is now viewed as a profession fit for women. The remaining problems within the occupation relate to improving the representation of women in senior posts and in maintenance and design functions but action is still taking place on these issues.

Lynne Segal, in her analysis of women's groups in the 1970s and early 1980s, has commented on our tendency to 'fail to take in and value our victories or to assess the strength and nature of the forces which determine our defeats' (Segal, 1987: 214). Using the spirit of this quotation rather than the words, this discussion will examine the strengths and weaknesses of the strategies used by the organised women's groups in the Institute of Housing and NFHA in the period 1980 to 1992.

Insecurity of the gains made

The gains made in the 1980s often rested upon the arguments that employers and the professional organisations needed to recruit and retain women as part of the pool of talent to staff their organisations. With the increasing recession of the 1990s there have been considerable fears that history will repeat itself and women be forced back into the home once labour is in over-supply. Although there seemed to be some evidence that this might be happening generally in employment, recent reviews have argued that women's employment may be maintained, although much of it is in low-paid and part-time work (Rubery, 1988; McGwire, 1993). Nevertheless, cuts in welfare provision affect many working women (Walker, 1988).

In housing work the threat to employment has also come from government policy. The funding framework for housing associations introduced by 1992 and the privatisation of local authority housing by such means as 'tenants' choice', large-scale voluntary transfers and compulsory competitive tendering for housing management represented a major shift away from the traditional view of social housing and towards a heavy emphasis on financial priorities and 'housing as a business'. Bearing in mind the stereotyping processes explored in Chapters 9, 10 and 11, fears that this shift of emphasis might disadvantage women are well founded. There was great concern that the new regime would have adverse effects on women as tenants and those seeking help with housing (Sexty, 1990). By 1993 there were fears that what was in hand was the elimination of social housing and of housing management as a profession (Institute of Housing, 1993; *Housing* editor, 1993).

Those working for equality were having to adapt their tactics to the new regime – for example a major focus in 1992 was to ensure that the Housing Corporation included equal opportunities in its perform-ance indicators. This was successful and its implementation remained a hopeful strategy for the future. But the contradictions of working within the financial framework imposed were having an adverse affect on the morale of many housing staff. Some were leaving if they could, and it remains to be seen whether this will affect women more than men.

Slow rate of underlying change

As well as the medium-term trends of recession and government housing policy, other longer-term trends were also a source of weak-ness. Previous studies have identified two major sources of disadvant-age for women in employment – stereotyping which may restrict their access to certain types of jobs and the reality of having the respon-sibility for 'caring' in society. Access to a wider range of occupations does seem to be improving but numerous surveys have indicated that it is still women who bear the day to day brunt of caring work. Although more lip service is paid to sharing of chores in the home, other factors, such as the government's 'community care' policy, may make the position worse (George, 1992). The efforts needed to change stereotypes clearly have to be maintained in the long term. Education and the media are quite rightly seen as key areas where the struggle has to be sustained. Women working in housing have ample cause to support these campaigns, while more equal distribution of jobs within the housing service will itself contribute to the battle against stereotyping.

Reliance on professional and employer-led bodies

The reliance on work with the Institute of Housing and the NFHA was a major source of criticism from some feminists. In particular the

Institute of Housing and professionalism were regarded with suspicion. There was quite widespread and genuine concern about the 'Queen Bee' syndrome – the fear that getting more women into senior positions would not necessarily help women who were disadvantaged. On the other hand attention to the professions need not mean ignoring the needs of women at other levels. The previous sections have illustrated how the campaigning on return to work and education by the IOH women's groups included the concerns of women at all levels. Most women who were members of the official working parties were actively involved with other groups concerned with women's participation as tenants or the impact of housing policy on women. They brought those issues to the NFHA and IOH working parties. The NFHA Standing Group was particularly active in this respect and made efforts to include such groups in its representation. The women's conference usually covered a wide range of such topics. Yet this produced more diversity of views within the working party and problems in this working party's relationship with the bureaucratic structures of the NFHA (for example it was not consulted about the Women at the Top report and was rarely free of the threat of being wound up). By 1993 the future of the NFHA standing groups and even its Equal Opportunities Sub-Committee looked extremely insecure.

The Institute of Housing Women's Working Party saw its role as ensuring that information about women's issues got through to members in senior positions with housing organisations. Where possible, dialogue was opened up with women in senior positions and they were provided with information. Some of them shifted to being more supportive of organised women's action. This type of change is consistent with the theories about behaviour of individuals belonging to disadvantaged groups (Tajfel, 1981: 284–286).

The fear that improving the position of professional women would not improve the lot of the majority remained a live issue to the end of the period. Public debate about the responsibility of women who were getting into powerful positions to promote equality for others (Kelly, 1992) and the evidence that, when more women get through, the 'Queen Bee' factor is not so powerful (Cooper and Davidson, 1982) maintained hope for the future.

The major strength of the strategy of working within the Institute of Housing and the National Federation of Housing Associations was that it made it easier to change both policy and practice, especially by the end of the period. Promotion of equal opportunities within the institutions themselves could have an immediate impact and this set an example to other employers. The opportunity to comment on professional practice, advisory publications and consultative documents was of considerable importance. Though neither organisation could change government policy there were whole areas of implementation over which they did have influence. The changes in education provide a good example of the way in which a network of decisions made over a long period of time were crucial to improving access for disadvantaged groups. From within the power structure it was possible to identify, at least on some occasions, where and when key decisions were going to be taken and to target pressure on that point.

Race, housing and feminism

This issue is an extremely important one because of the long and shameful catalogue of studies which have demonstrated direct and indirect racial discrimination in housing, starting with the PEP's *Racial minorities and public housing* (Smith and Whalley, 1975) and continuing into the 1980s and 1990s with adverse findings on local authority housing departments (for example, Commission for Racial Equality, 1984), housing associations (Runnymede Trust, 1985), estate agents (Commission for Racial Equality, 1990b) and building societes (Karn, Kemeny and Williams, 1985).

The Institute of Housing was slow to make an adequate response to this; it was not until 1985 that it produced a guide to race and housing monitoring written by its Race and Housing Working Party. The Institute did participate in the Path scheme (from 1984) to improve the flow of ethnic minority recruits into housing. In 1985 there was a decision to set up an ongoing working party on race and housing but nothing came of it until the establishment of its Equal Opportunities Working Group in 1990. Even this was intended to have only a limited life. Since black people were still under-represented within the Institute the attempt was made to improve this by co-option onto the Equal Opportunities Working Group. But the Federation of Black Housing Associations felt that the Institute's equal opportunities mission statement should go further to redress past imbalances by identifying and adopting positive action, and that unless the membership of the sub-committee was more varied in ethnicity and gender its effectiveness and credibility would be diminished (Julienne, 1991).

The NFHA had also come under criticism as being white dominated. It had a standing group dealing with race and housing, had an equal opportunities policy, and in 1986 had produced guidance notes on equal opportunities practice for speakers. Publications on race and housing appeared in 1987 and 1989 (NFHA, 1987; NFHA, 1989). The early women's groups at NFHA were mainly white. As we have seen, criticisms about exclusivity surfaced in the confrontation at the NFHA women's conference in 1989 and from then onwards the NFHA standing group made particular efforts to improve representation. This proved more successful in the case of conferences than with continued membership of the standing group.

During the 1980s the Federation of Black Housing Organisations was becoming more established and influential and some local authorities and housing associations were adopting more positive policies on race and housing at the local level. One organisation which achieved some success in improving representation was the Greater London Council (GLC). It had sufficient resources to be able to give some attention to black women within the remit of its women's equality work and publications (for example, GLC Women's Committee, 1986). The Housing Corporation formulated a strategy for black and ethnic minority housing associations in 1986 and developed this over the years, but by the 1990s the strategy had to be revised and there was concern that black ethnic minority associations would be adversely affected by the new financial strategies (Housing Corporation, 1993; Todd and Karn, 1993; Chandran, 1993).

Should either of the women's working parties have done more to promote race and housing issues? Bearing in mind their relative weakness in the power structures what should they have done? Probably more could have been done to ensure that black women were able to participate in the discussions. More attention might have been drawn to race issues and to black women and housing in briefing and discussion papers. In the early stages few black women were members of the Institute of Housing. All the working parties found that women had difficulty in attending regularly because they were in more junior positions at work and often had domestic commitments. Black women were even more likely to be at lower levels in the organisation. In these circumstances the tactics used (mostly by the NFHA) included making sure that fares to meetings were paid for those working for organisations with few funds, trying to ensure that times and days of meetings were convenient, seeking out groups and contacts, ensuring the circulation of minutes and papers to those unable to get to meetings and attempting to keep networking and consultation with unrepresented groups alive. The IOH women's working party did consistently raise such issues as comprehensive equal opportunities procedures and monitoring. But it could be seen as inappropriate to raise black women's issues without black women participating. Both working parties were likely to be challenged if they were considered to have overstepped their remit.

By 1984 a Black Women and Housing Group had formed (Welch, 1986). The first report on black women and housing was published in 1990 (Rao, 1990). But given the lack of resources it was difficult for such groups to maintain the long-term presence which was needed to make an impact on bureaucratic organisations.

In the 1990s the relationship between feminism and race equality remained an issue both for black women and white women, inside and outside of the housing field (for example, Adams, 1989; Harriss, 1989; Solomos, 1991; Rao, 1990; Jayaweera, 1993; Peach and Byron, 1993). Black women felt that even where there were official working parties these usually had separate sub-groups on race and women and their needs were not properly dealt with by either.

Strategies and tactics for equality work

This study confirms the importance of many well established strategies and tactics. For example the formulation of an equal opportunities policy by the NFHA and the Institute of Housing was by no means enough in itself but was an essential first step since it provided a base for many subsequent discussions. The working out of equal opportunities procedures and guidance, like the NFHA's Code of Guidance to speakers, is an important subsequent stage which may take some time (and may need to be repeated because staff change and organisations can 'forget'). The keeping of monitoring statistics is essential but they are only going to be effective if they are reviewed and used. Both institutions at times kept statistics without presenting them, or presented them in such a way that they were not easy for a lay committee to use. But the same statistics could be invaluable when arguing a case for change.

Special surveys to examine discrimination in employment may have to be repeated at intervals as assumptions may be made about improvement which are not justified. On the other hand effective use of monitoring data can reduce the reliance on surveys and reduce cost. By the end of the period the setting of specific targets had become more common but these also need monitoring and action – the NFHA Conference had accepted a target of 50 per cent women speakers at conferences in 1985 but this was never monitored and seems to have become a dead letter.

Alliances

To bring sufficient pressure to bear, an alliance of groups with different interests was usually essential; for example between women with an established and long-term interest in equality and women who were of more conservative orientation but were motivated by personal experience of discrimination. Since men were in a majority on both Councils, getting some male support was essential. In both Institutions there were times when effective action was dependent on one or two sympathetic Council members (male and female) who were willing to present a case. As in most bureaucratic institutions of this type, staff support was also very useful in promoting change.

Such alliance could be somewhat fragile and easily rent apart by differing theoretical views and interests (Segal, 1987). Concentration on goals and specific changes being sought is useful when it is appropriate to keep such an alliance together. At an early stage the Women in Housing Group decided that, given the shortage of resources, it was essential for each group to decide 'what is it that this group can do better than other existing groups?' This remained a key question into the 1990s. If each group concentrated on its own task, it needed also to recognise and respect the functions of other groups working on related tasks. If respect and tolerance for divergence of views was maintained then viable alliances could be formed to pursue specific issues while maintaining each group's identity and values.

Persistence and support

One of the key points illuminated by this study is the length of time and perseverance needed to bring about change at this level. In both the NFHA and the IOH, improvements were at one point agreed but later forgotten only to be debated again a few years later. Both working parties were at times threatened with dissolution and often criticised simultaneously for doing too much and too little.

To carry on over a long period of time these groupings need to be continually renewed and supported. This may be a daunting prospect. The Women in Housing Group successfully used the tactic of concentrating first on issues where it would be relatively easy to make headway with the rest of the profession. Success tends to help to build groups. But in the long term more difficult issues need to be tackled as well.

Underpinning this long-term work must be formal or informal networks of women who have clear ideas about the outcomes they want for women's equality. A profession can provide a stable national structure which may help women form such networks. It may be that one of the advantages housing and planning have over architecture and surveying is that they are relatively small professions – it is easier for small groups of women with few resources both to contact each other and to influence the majority. But, as we have seen, it was essential that the networks did not begin or remain within the professional groups – continued input both from groups concerned with the impact of housing policy on women and with wider work for women's equality in employment was vital for supporting and renewing this struggle.

The attenuating nature of such long-term struggles is evident. Workshops at the 1991 and 1992 NFHA Women's Conferences on 'working in male-dominated organisations' confirmed that many women often felt worn down by the continual challenges they faced. They needed better resources both for personal support and for negotiating within organisations. One important resource which has been crucial for many women is supportive contact with other like-minded women. But some women either had no suitable group within their locality, or were put off by an aggressive or dogmatic style in a particular group. While local support is ideal, national and regional networks and newsletters can play a vital part here. Other women might feel that, having a working life and domestic commitments, they did not have time for such activities. The recognition that maintaining their own well-being is crucial, both to caring for others and to campaigning, comes slowly to many women. There are resources already available from areas such as negotiation skills, personal development and group dynamics which could help to sustain women and make them more effective in the long-term struggle. The opening up of such possibilities remains important for the future.

Flexibility

A realistic and pragmatic approach, demonstrating positive achievements as well as criticising existing policy, perhaps suited the style of women with backgrounds in housing management and other forms of housing practice. It required flexibility. For example both groups at times faced a demand to have male members. While more dogmatic groups might have objected, these groups decided that it was appropriate to agree in the knowledge that they could maintain all-women groups elsewhere. It was fascinating to see that when men were in a very small minority they rapidly stopped attending.

The groups on the whole followed tactics of negotiation rather than confrontation. This was important as it provided a bridge to the uncommitted, who were more easily won over by persuasion rather than confrontation. But if carried to excess it could lead to stagnation and a fear of 'rocking the boat'. Too much flexibility can lead to accusations of a lack of principle (Kelly, 1992). So there were times when assertion of principles and confrontation were appropriate and important – for example at the 1989 NFHA Women's Conference and within the Institute of Housing in 1989–90. It is useful to note that in both cases the issue chosen for confrontation was a substantial and long-term one and as soon as confrontation occurred attention was given to continuing dialogue and seeking grounds for further progress. 'Making change involves challenges, as well as access to forms of power.' The four strategies for change which Kelly (1992) identifies:

– raising issues, challenging definitions;
– creating alternative institutions;
– gaining status and influence within institutions;
– calculating campaigns/using forms of direct action,

were all used to create the changes of the 1980s. In meeting the challenges of the 1990s it seems likely that women in housing will need all the diversity of strategies and resources we have identified so far, plus new ones to meet new situations.

12 Conclusions

OVERVIEW

Octavia Hill had popularised the idea of good housing management and of employing women to do it. From 1912 onwards women who had been trained by Octavia Hill began to gather in small groups and by 1932 these groups had managed to overcome personal rivalries and differences of outlook and formed one society for trained women working in housing management.

In the period 1932 to 1939 state involvement in housing provision expanded rapidly and the opportunities for women's employment increased. But the opposition, based on the fact that this employment did not fit neatly into the dominant stereotype of the female role, also grew. Even when the employment of women was considered acceptable they could be relegated to a narrow 'welfare role' and it was usually unacceptable for them to manage male staff. The Institute of Housing, set up by predominantly male local authority staff in housing and dominated by men, seems to have supported this narrow interpretation of women's proper place. The women's Society therefore played an essential role by recruiting and training women, facilitating strong links between them and providing the social support which would enable them to survive difficult or isolated situations. The Society's development of the role of the woman housing manager, who had substantial overall responsibility for estates in her charge, helped to give these women, who were breaking through convention, a clear professional identity. Through publicity, lobbying and negotiation the Society was able to encourage at least a proportion of the major housing authorities to employ women as housing managers. It linked with other women's organisations when necessary to put pressure on government about general housing issues. On the other hand, the strictness of the Society's training system kept it small in size and middle class based. Their argument that women had a 'special aptitude' for housing work was itself based on a stereotype and could be turned against them. Many of the women pioneers were single women who felt that having a career ruled out marriage and children. Some were married and some did manage to combine career and children, relying on the availability of paid help. Others found they needed similar help for looking after elderly and sick relatives or had to relinquish careers in order to do this. As education spread, housing work began to draw new recruits from the less well-off but the lack of public money for training was always a barrier to wider recruitment. Nevertheless it was the basis of employment, laid in the pre-war period, which enabled women to advance further into the traditionally male-dominated world of housing during the Second World War.

The war years, which have been almost ignored by all the existing histories of housing, were years of development of women's employment. They proved that women could do difficult and dangerous work in housing and elsewhere. Women were employed in a much wider range of housing jobs, including posts with Central Government, and were considerably involved in plans for reconstruction after the war. Society members were, however, aware that the gains might not be sustained after the war despite the hopes for a brave new world.

After 1945 public housing expanded rapidly. Though some women were pushed out of housing work, the expansion of opportunities meant that most of those drawn in by the war continued successful careers. But the expansion of public and social housing began to mean that jobs were more attractive to men. The initial spurt of new local authority building after the war slowed down in the 1950s and 1960s, especially under Conservative governments. Local authorities still played a substantial role in slum clearance and rehabilitation and from 1964 housing associations began to grow. In the 1950s and 1960s local authority stocks grew larger, and by the late 1960s local authorities were expected to take on a more comprehensive role in housing. There was still no public investment in training for housing work and the housing practices of many departments and associations were increasingly criticised as paternalistic and discriminatory. The restrictions in the Society's training scheme meant that it did not keep pace with this new demand. Despite social changes towards greater equality for women, child care and dependency continued to be problematic for women who wanted to stay in employment, and the availability of paid or communal help declined. At the same time the social climate produced difficulties for women who did not marry or have children.

It was during this period that some of the disadvantages to women in housing of having a separate organisation became more apparent. The separation of men and women played an important part when people were appointed to senior jobs. If they were in Society offices they made a restricted range of career contacts but were not so well known to senior Institute members and might miss out on promotion opportunities. Women in smaller housing trust offices might be adversely affected by this. This factor was less likely to affect women in local authority employment, who tended to have working relationships with Institute senior officers in their areas. But female housing trainees who were in non-Society offices might lack the kind of support which the Society gave, though some did receive this from enlightened senior staff. The commitment of the Society to a particular type of housing management which stressed the welfare of individual tenants was also likely to have reinforced gender stereotyping. Women were seen as concerned with the individual, the small scale and the social side of housing. Consequently as housing grew and changed, Society members were labelled 'old-fashioned' and of limited vision. The 'housing is a business like any other business' approach worked against women's employment. There is irony in the fact that the progressive housing management of the 1980s and 1990s came back to stressing many of the principles for which the Society stood.

Despite the continued influence of stereotyping, the 'surface equality' of the 1960s meant that continuing to have a separate women's society was almost seen as illegitimate. There were strong financial and practical incentives for the Society to amalgamate with the Institute of Housing. The ethos of the time meant that no long-term safeguards for women's participation or women's interests were included in the amalgamation arrangements.

By 1972 there had been a dramatic reduction in women's participation within the professional organisation. Chapter 8 examined the way in which the wiping out of a separate women's society left their interests unprotected, and analysed the effects of group dynamics and sex role stereotyping within the new Institute. But these changes were taking place against a background of change in society and housing policy which also affected the employment of women. Chapter 9 identified the way in which the growing emphasis on large organisations adversely affected women's access to senior posts from 1945 onward. In contrast, equal opportunities legislation, changes in housing education and training and the expansion of different types of work, such as housing aid, gradually increased the number of women at lower levels. By the 1970s and 1980s employment surveys began to provide more comprehensive information on male and female employment in housing and to illustrate horizontal and vertical occupational segregation. By the 1980s substantial changes in government policy were residualising the housing service and aimed to transfer much of the remaining stock from local authorities to housing associations and the private sector.

In the 1980s not only did women's participation in the housing profession begin to rise again but women's organisations reappeared, albeit in a different form. Considerable improvements were achieved but new challenges have arisen in the 1990s. The changes in the role and ethos of social housing organisations may present the most difficult challenge for women since the 1970s.

The related professions of planning, architecture, surveying and accountancy, examined in Chapter 10, reveal a parallel pattern of disadvantage and a number of initiatives towards equal opportunities in the 1980s. There are encouraging signs that where professional organisations have taken action they have increased the proportion of women and that equal opportunities legislation is having some effect. The RICS was the slowest to take action and is the only one without active equal opportunities work in 1992. There is some evidence that the male stereotyping of landed professions is more difficult to shift than that of the financial professions and ample evidence that it is not yet time to be complacent about progress made.

LESSONS FROM THE PAST

The influence of the Society of Housing Managers

Care always needs to be taken with the lessons we draw from history (Tosh, 1984; Popper, 1957). But we all want to learn from our collective experience so that we can move on in the future. Given that hindsight is of limited usefulness, it is still interesting to speculate on two might-have-beens. Would women have had a significant place in housing work had there been no Society, and could women's participation have been safeguarded any better after unification?

If there had been no Society it seems unlikely that there would have been any substantial participation by women in housing work in the 1920s and 1930s. The housing departments and associations, which were outside the Society's influence, were staffed by men, often recruited from other local authority departments or from the armed forces. Where women were employed it was usually in a clerical or welfare capacity. Consequently there would then have been very few, if any, female housing managers and the opportunities offered by the Second World War could not have been taken up so extensively. Even after the war women's progress would have been limited. This can be demonstrated by examining women's progress in the Institute-dominated departments. A few of these departments began to recruit and train women by the 1960s and the number increased with changing educational patterns and with the advent of equal opportunities legislation. But even by 1968 there were still some all-male housing departments and very few Institute-trained women in senior positions.

On the other hand it could be argued that the subordinate position of women within the Institute from 1965 to the late 1970s was made worse by the prior existence of the Society. Many of the most talented and best-educated women who wanted to work in housing became Society members; unless they were remarkably clear about career patterns, and determined, they had little incentive to join the Institute. But when amalgamation came in 1965 this meant that some of these women were not well prepared to face the politics of operating within the new Institute. This study has recorded in detail the effects of this. It appears that the position of women within the new Institute would have been better safeguarded if some kind of recognised women's grouping continued, but at the time senior women were committed to making the amalgamation work and felt that such a course of action would be divisive.

By the 1980s social change and equal opportunities legislation had produced a different climate. Nevertheless, this study has shown that awareness of women's previous prominence in housing was a resource in the renewed struggle for equality. The achievements of the Society of Housing Managers and the fact that women had occupied senior positions in housing employment could be contrasted with the low participation of women in the Institute and in higher-grade employment in the 1970s and 1980s. Women in accountancy, planning and surveying lacked this resource. There had been individual female achievers in these occupations but younger women did not have the same kind of historical group presence to refer to. Greed has commented on the difficulties arising from the lack of a sufficiently strong identity for the woman surveyor. There is a 'chicken and egg' argument here with regard to role models: it is only when there is a sufficiently large group of senior women with varied backgrounds and outlook that younger women are likely to find a role model compatible with their own ideas. Though women's public participation in housing work reduced sharply following the 1965 amalgamation, a large group of women remained in housing and there was a slightly higher proportion occupying middle and senior management posts than in the related professions. There were also some 'enlightened' men in senior positions who provided key help in the early battles for recognition of women's issues.

The diversity of women's organisation

Arguments about the desirability or otherwise of separatism are as old as the women's movement. This study indicates that in housing all-women groups were needed at a number of different stages, but that they took many different forms. Informal networks, small pressure groups, formal working parties and sub-groups as well as a women's professional organisation and a recruitment and training agency were used. As Connell (1987) has noted, there was a need to form alliances with other special interest groups and groups which included men. The advisability of forming such alliances has been for many years a subject of debate within the women's movement (Segal, 1987: 56). Doubts about women in positions of power form a counterpoint to this study (Eisenstein, 1990; Rowbotham *et al.*, 1979). We can recognise the hazards of such alliances but remain convinced that they are essential in order to bring about the long-term changes required for women to be free to express their potential. Flexibility in setting up new forms of organisation or in allowing them to disband was useful.

Male power and the persistence of stereotyping and discrimination

Women in discussion groups often express concern about the persistence of male power and stereotyping despite the efforts made to counteract it. The 80 years of women working in housing illustrate this only too well. In the 1930s married women were told that they could not do housing work because they needed to care for home and husband, and unmarried women were told that they lacked broad enough experience of life. In the 1990s younger qualified women may be told they lack experience while being blocked from getting it, while older women returning after a career break may be told they are too experienced for a first-line management job. Even in 1992 a woman had to take a district council to court in order to get equal pay for equal work (*Inside Housing* editor, 1992). The logic of arguments does not seem to matter too much provided that they keep women in their place. The pervasive nature of male power and recurrence of the same battles may seem depressing, especially at present when both in our country and elsewhere inequalities are multiplying. One woman likened sexism to the science fiction idea of an animal which can constantly change its shape and appearance to adapt to changing circumstances.

There are some long-term changes which are favourable to feminism; advances in contraception and medical care mean that child care does not claim as large a place in women's life as it did previously. On the other hand the claims of dependency care are growing. A number of women of the pre-war generation in housing had their careers affected by the need to care for sick or elderly relatives and there is ample evidence that this is still primarily seen as a woman's role. It is important for women that the need for and value of this work is stressed but also that it does not fall solely on women. This will require further campaigning.

Women have been oppressed in our culture for thousands of years. The length of time it takes to change embedded habits which underpin the structures of power is unwelcome, but not surprising. If we can use this awareness to encourage us to make our campaigns more effective and support those who are involved then it will be a source of strength rather than weakness.

AGENDA FOR THE FUTURE

Continuance of action within the professional organisations

This study has confirmed the importance of positive action to open up all types and levels of work to women. It has demonstrated that professional organisations can play a constructive part in this and that their action can also be a help to those working below professional level. It is not yet safe to leave this issue to chance. The main focus for action in the 1990s is to ensure that women, once recruited, can progress to senior jobs and the professional Council. But the continued discrimination against women in the technical functions of housing has a common cause with discrimination against women surveyors and architects, and co-ordinated action on this issue is likely to be useful. It is important that equal opportunities action be confirmed and continued both within housing and in related professions. Effective action depends both on the right policies and on implementation. Chapter 10 spells out the detail of what is required. Similar action is needed in relation to other powerful institutions, such as the judiciary, parliament and the management of large organisations.

Broader social action

Women's participation in housing work began because of a concern for social justice and a desire to ensure suitable housing for all the population. This concern has been continued in the 1980s with attention to the needs of particular groups and to issues of class, race, age and disability. There has also been a slowly growing awareness of women and shelter issues world-wide. Issues of poverty and of who does the caring are particularly important in relation to women's housing needs. These link women in housing with many other groups which continue to state the case for social justice. Paying attention to employment in the housing service and to individual staff support needs to be seen in the context of this broader social action. Details are not spelt out here as they are well covered in other texts (for example, Segal, 1987; Gilroy and Woods, 1994). But policies which have the effect of undermining the whole future of public and social housing are a clear threat to women and to many other disadvantaged groups.

Changing the stereotypes

Language not only expresses what we think but also structures the way in which we think. Feminists have quite rightly made stereotyped words and images in education and the media a prime target over the last 20 years and this continues to be crucial. Women in 'non-traditional' occupations such as housing or surveying help to combat stereotyping just by being there. In addition, if they seek to improve equal opportunities practice within the profession they open up opportunities for others to demonstrate that it is not just exceptional women who can do these jobs. It is also vital to challenge stereotyping of these occupations in the media and to identify how much this restricts women's career opportunities. Equal opportunities policies and practice in the educational system are central to reducing the remaining areas of discrimination relating to building work and to management. We must challenge the stereotype that 'caring work' is uniquely part of women's role and establish that women are successful in technical or senior management roles.

Support for women

It was argued in Chapter 11 that women, especially those currently engaged in campaigns for equality, need to pay more attention to their own support and well-being. There are all kinds of reasons why women may not pay sufficient attention to this. Early childhood conditioning to look after other people's needs may be translated into a dedication to 'the cause' which inhibits care for personal well-being. The deep-rooted feelings of lack of self-worth which our society engenders in many women may prevent even those women who seem successful from caring for themselves. Even when these problems are overcome women may not be aware of the kinds of support they need or how to get that support. The threats to women's equality inherent in present policies, the evidence that the struggle is a long one and subject to reversals, can be transformed into reminders that we need to care for ourselves and ensure that we have the support to keep the struggle going. The agenda for the future therefore includes maintaining support systems and developing the skills which will enable us to campaign effectively in future. Some aspects of support are outlined below but individual women may need more or less of each according to their circumstances.

Group support and identity issues

Formal and informal groups play a central part in campaigning and are an important source of support. Part of that support can be help in working out aspects of professional identity. Greed has described how for some women surveyors the image of the lady surveyor did not provide an acceptable professional identity. Women need a range of role models to choose from so that they can find someone who is compatible with their own outlook. Women in senior positions some-times find acute difficulty with being regarded as a role model. But when there is only a limited number of women in an occupation it is clear that some younger women are encouraged or discouraged by the example of senior women. Helping senior women deal with this role appropriately is an aspect of support which has been neglected. For some women the suitable reference group may refer to other aspects of identity; for example, the Older Feminist Network does unique work in seeking to change the stereotype of older women.

Shared values and the development of theory

Those who care about equality can find themselves challenged every day of their working lives. This is wearing and it is essential that there is contact with others who share the same values, however that contact is maintained. Local contacts are ideal but for those who lack them national networks and newsletters are essential. A major function of such groups is to help the individual feel that she is not alone in facing challenge and to work out successful strategies for creating change. Provided that there is sufficient commonality of experience, this function can be fulfilled by many different types of group.

It has been a cornerstone of the women's movement that women come to an awareness of inequality through their own experience. However, beliefs which are based solely on one's own experience may fail to provide sufficient support in a challenging world. So from its early days 'consciousness raising' has included broadening out from the individual to the shared experience. The development of theory has not stood still. Discussion, argument and debate has continued throughout the 'death of the women's movement', post-feminism and any other labels the media choose to invent. It is an essential part of maintaining a living and growing movement. Critics may see debate as a weakness, while others recognise it as a sign of strength and openness. Individuals need to develop their own theoretical understandings so that they have coherent and well evidenced arguments with which to deal with opposition. For those who are involved in campaigning, theory is not a luxury but an essential.

Human warmth and social relationships

In recent years there has been a reaffirmation of traditional 'feminine values' which lay a stress on the importance of intimate personal relationships. We can acknowledge the importance of these values without falling into the trap of assuming that they are innately either male or female (Segal, 1987: 1–37). Modern psychology has confirmed how significant these are to the health of the personality for men and women. Close relationships are essential to anyone's well-being, and need time. This study has confirmed that social contacts made through the occupational network can play an important part in providing that personal support but for most people family and friends will be the other major source. Nevertheless, much working life, particularly (but not only) in senior jobs, is still organised as though an individual's whole time can be dedicated to work. The recession seems to have increased the tendency for some organisations to run a culture where those who seek promotion are expected to work excessive hours. There is a clear conflict in government policy, which pays lip service to the family and yet removes restrictions on working hours. To seek to change this tendency may seem idealistic at the present time, but the vision of what we want must be established so that we can begin to work towards it. The fundamental shifts needed to bring about a more equitable distribution of paid work and caring work between men and women and to allow the time for satisfactory relationships are within the capabilities of the world we inhabit. They can and should form part of the growing concern for ecological development. In the meantime women can be supported in seeking to maintain a suitable balance in their own lives.

Training for campaigning

Women campaigning in housing demonstrated the ability to identify key issues, assess the strength of opposing forces and decide on suitable tactics. These skills are often developed *ad hoc*, but many different kinds of training could enhance existing skill levels and enable women to create change more effectively. For example, women often still regard negotiation skills as something that belongs to management or unions and do not see the potential of such courses for themselves. They may have justifiable doubts about the kinds of values they might encounter on courses for negotiation or other business skills. Yet bringing about change in organisations or society usually requires negotiation. Clear thinking about desired outcomes would enable women to make better use of existing courses and resources. It would be even better if more mainstream business courses demonstrated greater awareness of equal opportunities values and thus were more acceptable to women.

Similarly, most feminists were for years very wary of the kind of personal support available from psychological approaches. This originally arose from justifiable reservations about much theory and practice arising from psychology and psychiatry (Chesler, 1973; Franks and Rothblum, 1983). But 'a lot of women need individual help to be able to fight, or even to keep going' (Pixner, 1978). The human potential movement, personal development and new approaches based on learning theory (see for example Laborde, 1987 and Watzlawick *et al.*, 1974) have broadened the range of what is available. Women have been benefiting from assertion courses in the 1980s and these are still needed. But 'beyond asser-tion' towards a broader range of suitable personal development is an appropriate direction for the 1990s. A woman who has resolved internal conflicts and become more integrated, in control of her anger rather than controlled by it, becomes stronger and more effec-tive in creating change. Groups which have the ability to face and resolve internal conflicts and to continue to develop can become more effective in influencing the external world.

CONCLUSION

After 80 years there are still women working at all levels of the housing service because they have a commitment to the provision of housing and to social justice. There are more of them now. The support mechanisms available to a relatively small group of women through the Society of Housing Managers have been replaced by wider and more diffuse networks of groups and working parties. A broader range of support and training in relevant skills is also avail-able for those who know how to use it. The challenges of the 1990s are such that continued flexibility in organisation, support, campaign-ing and negotiation skills will be needed to ensure that women are fully able to participate in the policies and practice of social housing work in the twenty-first century.

Appendix: Sample interview schedule

CAREER

When did you first start in housing work?
Why did you choose this career?

At which offices did you train?
What kind of training did you have?
What kind of housing were those offices dealing with?

What was your first appointment?
How did you hear about it?

What kind of responsibilities did the housing department (organisation) have?
What kind of responsibilities did you have?
What kind of housing and tenants was it dealing with?
Are there any particularly interesting aspects of its housing management work?

(*These questions are repeated for each subsequent job.*)

Is there any person or persons whom you regard as having being influential in your career?

Do you think that being a woman affected your career in any way?

THE SOCIETY OF HOUSING MANAGERS

What was your first contact with the Society? What impression did it make on you?

Did you take part in its Council, committees, etc.? What kind of work did you do?

Why did the Society decide upon unification with the Institute of Housing?

What did you see as having been the gains and losses of unification?

Did women participate in the new Institute? What factors affected this?

(If the content of later questions had already been covered in answers to earlier ones they were not repeated.)

(From *Inside Housing*, 6 April 1984; published by the Institute of Housing)

Bibliography and references

Adams, M.L. (1989). There's no place like home: on the place of identity in feminist politics. *Feminist Review*, 31 (Spring): 22–23.

Adams, C. and Laurikietis, R. (1976). *The gender trap*. Book 1: *Education and work*. London: Virago.

Adamson, O., Brown, C., Harrison, J. and Price, J. (1976). Women's oppression under capitalism. *Revolutionary Communist* No. 5.

Adorno, T.W., Frenkel-Brunswik, E., Levinson, D.J. and Sanford, R.N. (1950). *The authoritarian personality*. New York: Harper and Brothers.

'AFL' (1942). Annual general meeting, 16th November 1941. *SWHM Quarterly Bulletin*, 35 (January): 3–6.

Agate, W.J. (1936). Letter from Wm Agate as secretary of the Peabody Trust, to the secretary of the Balfour Sub-Committee. PRO HLG 37/5.

Alford, H. (1942). Letter to the editor. *SWHM Bulletin*, 37 (September): 8.

——— (1952). 100 years of weekly property. *Society of Housing Managers Quarterly Bulletin*, 3 (6), April: 4–8.

——— (1981). Interview with M. Brion, 27th February.

Alford, H. and Upcott, J.M. (1933). Letter dated 20th October 1933 re handover of funds from AWHPM to SWHEM: Miss Upcott's papers, held by M. Brion.

Allen, J. (1990). Review. *Roof*, March/April.

Allen, L. and Lawes Wilkinson, K. (1928). The Cumberland Market Estate. *Octavia Hill Club Quarterly*, December: 3–5.

Allen, P. (1981). *The role of housing associations in Britain*. Unpublished dissertation for MPhil, University of Bristol.

Allport, G.W. (1958). *The nature of prejudice*. New York: Doubleday Anchor Books.

A Member of the Women House Property Management Association (1919). *House property management. Miss Octavia Hill's system*. London: AWHPM. Reprinted from the *Land Union Journal*, June 1919.

A Mere Male (1944). Letter to the editor. *SWHM, Bulletin*, 41 (June): 12.

Anon. (1911). A Manchester experiment. Reprinted from the *Manchester City News*, 21st & 28th January 1911. Miss Upcott's papers. Held by M. Brion.

——— (1931). Women as estate managers. *The Times*, 9th February. Pamphlet reprinted and published by the Times Publishing Company, London.

Argyle, M. (1967). *The psychology of interpersonal behaviour.* Harmonds-worth: Penguin.

Ashridge Management College (1980). *Employee potential – issues in the development of women.* London: Institute of Personnel Management.

Association of Municipal Corporations (1937). Memorandum on the management of municipally owned houses. PRO HLG 37/5 Paper B22.

Association of Women Housing Workers (1916a). *Constitution, council, committee and membership leaflet.* London: Association of Women Housing Workers, later AWHPM.

—— (1916b). *Working-class houses under ladies' management.* London: Association of Women Housing Workers.

AWHPM (1917a). *Constitution, council, committee & members.* Dated by hand by Miss Upcott 1917. Published at Women's Institute. London: AWHPM. Miss Upcott's papers held by M. Brion.

—— (1917b). Annual Report for the year 1916. Marked by hand by Miss Upcott 1916. First Report but publication date likely to be 1917 because it notes officers for 1917. London: AWHPM.

—— (1918a). *Constitution, council, committee and members.* Dated by hand by Miss Upcott 1918. London: AWHPM.

—— (1918b). Annual Report for 1917. London: AWHPM.

—— (1919a). *Constitution, council, committee & members.* Dated by hand by Miss Upcott 1919. London: AWHPM.

—— (1919b). Annual Report for 1918. Updated typescript likely to be 1919. London: AWHPM.

—— (1920a). *Constitution, council, executive committee, details of training scheme.* Dated by hand by Miss Upcott 1920. London: AWHPM.

—— (1920b). Report of council meeting 7th October 1920. Duplicated typescript, damaged. London: AWHPM.

—— (1921). Annual Report for 1920. Duplicated typescript. London: AWHPM.

—— (1921–1926). *House property management by trained women.* No date. Originally published from 48 Dover Street, but re-used at 3 Bedford Square. Dated from this and internal evidence as 1921–1926. London: AWHPM.

—— (1927). Annual Report for 1926. Duplicated typescript undated but likely to be 1927. London: AWHPM.

—— (1930a). *Council and association membership.* Printed leaflet, undated. London: AWHPM.

—— (1930b). Annual Report for 1929. London: AWHPM.

—— (1930c). *Visit to Holland.* Duplicated paper. London: AWHPM.

—— (1931a). *Council, committee & members.* London: AWHPM.

—— (1931b). Annual Report for 1930. London: AWHPM.

—— (1931c). *Financial aspect of house property management.* London: AWHPM.

—— (1931d). *Training scheme.* London: AWHPM.

—— (1932). Annual Report for 1931. London: AWHPM.

—— (1933). Annual Report for 1932. London: AWHPM.

Balchin, P.N. (1981). *Housing policy and housing needs.* London: Macmillan.

Balfour Sub-Committee (1935). Minutes of first meeting, 22nd April. PRO HLG 37/4.

———— (1936). Minutes of third meeting, 30th July. PRO HLG 37/4.

———— (1937). Minutes of seventh meeting, 16th April. PRO HLG 37/4.

———— (1938). Minutes of meeting, 27th January. PRO HLG 37/6 Paper B 37.

Banaka, W. (1971). *Training in depth interviewing*. London: Harper & Row.

Barclay, I. (n.d.). *Property management*. An essay from *The Road to Success*, ed. Margaret Cole. London: Methuen for SWHEM.

———— (1976). *People need roots. The story of the St. Pancras Housing Association*. London: Bedford Square Press.

———— (1981). Interview with M. Brion, 25th March.

Barron, R.D. and Morris, G.M. (1976). Sexual divisions and the dual labour market. In Allen, S. (ed.) *Dependence and exploitation in work and marriage*. London: Longman.

Barrow, M. (1980). *Women 1870–1928*. London: Mansell Information Publishing.

Baskett, M.R. (1962). Early days in Liverpool. *Society of Housing Managers Quarterly Bulletin*, 5 (6), April: 7–9.

Baynes, A. (1935). An experiment in rehousing. *SWHEM. Quarterly Bulletin*, 8 (January).

Beechey, V. (1978). Women and production. In Kuhn, A. and Wolpe, A.M. (eds) *Feminism and materialism*. London: Routledge and Kegan Paul.

Beddoe, D. (1983). *Discovering women's history. A practical manual*. London: Pandora.

Bennett, W.S. and Hokenstad, M.C. (1973). *Full-time people workers and conceptions of the professional*. Keele: University of Keele.

Berry, F. (1974). *Housing: the great British failure*. London: Charles Knight.

Binney, V., Harkell, G. and Nixon, J. (1981). *Leaving violent men*. Manchester: Women's Aid Federation.

Birmingham Feminist History Group (1979). Feminism as femininity in the 1950s? *Feminist Review*, 3: 48–65.

Blair, F. (1993). *Single women in housing need*. London: NFHA.

Block, J.H. (1976). Issues, problems and pitfalls in assessing sex differences. *Merrill Palmer Quarterly*, 22: 283–308. Referred to in Griffiths & Saraga (1979).

Blyth, O.M. (1933). Housing in Western Germany. *SWHEM. Quarterly Bulletin*, 3 (October).

———— (1934). Working class dwellings in and around Paris. *SWHEM. Quarterly Bulletin*, 6 (July).

Bock, R.D. and Kolakowski, D. (1973). Further evidence of sex-linked major-gene influence on human spatial visualizing ability. *American Journal of Human Genetics*, 25: 1–14.

Bouchier, D. (1983). *The feminist challenge*. London: Macmillan.

Bowley, M. (1945). *Housing and the state*. London: Allen & Unwin.

Boyd, N. (1982). *Josephine Butler, Octavia Hill, Florence Nightingale*. London: Macmillan.

Boyd, W. (ed.). (1944). *Evacuation in Scotland*. Bickley: University of London Press.

Brailey, M. (1987). *Women's access to council housing* (Occasional paper No. 25). Glasgow: The Planning Exchange.

Brenner, M., Marsh, P. and Brenner, M. (eds) (1978). *The social contexts of method*. London: Croom Helm.

Bridenthal, R., Koonz, C. and Stuard, S. (1987). *Becoming visible: women in European history*. Boston: Houghton Mifflin.

Brion, M. (1982). Women in housing. Action which the Institute needs to take. Paper presented to London Branch of Institute of Housing, 4th November.

———— (1984). Talk at Institute of Housing Conference at Harrogate, 29th June.

———— (1989). The Society of Housing Managers and Women's employment in housing. PhD thesis. London: The City University.

———— (1992). Women and anger. Personal development workshop for the Older Feminists' Network. Spring.

———— (1994). Snakes or ladders? Women and equal opportunities in education and training for housing. In Gilroy & Woods (1994).

Brion, M. and Tinker, A. (1980). *Women in housing, access and influence*. London: Housing Centre Trust.

Bristol Women's Studies Group (ed.) (1979). *Half the sky*. London: Virago.

Brown, A.J. (1936). The equal pay debates. *London Town*, 37 (437) May: 172, 3.

Brown, K.C. (1961). Fifteen years in North Kensington. *Society of Housing Managers Quarterly Journal*, 5 (1) January: 6.

Brown, M. (1936a). Slums and social service. *London Town*, 37 (433) January: 25, 26.

———— (1936b). Housing estate management. Letter to editor. *London Town*, 37 (435) March: 94, 95.

Bruegel, I. (1979). Women as a reserve army of Labour: a note on recent British experience. *Feminist Review*, 3: 12–23.

Burgess, R.G. (1985). *Strategies of educational research*. London: Falmer Press.

Burnett, J. (1978). *A social history of housing*. Newton Abbot: David and Charles.

———— (1986). *A social history of housing, 1815–1985*. Second edition. London: Methuen.

Burney, E. (1967). *Housing on trial*. London: Institute of Race Relations.

Byrne, E. (1978). *Women and education*. London: Tavistock.

Byrne, T. (1986). *Local government in Britain*. Harmondsworth: Penguin.

Calder, A. (1969). *The people's war: Britain 1939–45*. London: Jonathan Cape.

Cape Argus (1935). *This housing problem*. Cape Town: *Cape Argus*, 21st June.

Carey Penny, M. (1941). Romford. *SWHM Quarterly Bulletin*, 32 (January): 7–9.

CHAC (1938). *The management of municipal housing estates*. Report of the House Management and Housing Associations Sub-Committee of the Central Housing Advisory Committee. London: HMSO.

———— (1939). *The management of municipal housing estates*. Report of the House Management and Housing Association Sub-Committee of the Central Advisory Committee. London: HMSO.

—— (1942). CHAC Sub-committee on the design of dwellings. Minutes of second meeting 10th July. PRO HLG 37/62.

—— (1943a). CHAC Sub-committee on the design of dwellings. First meeting of panel on flat development, 1st February.

—— (1943b). CHAC Sub-committee on the design of dwellings, Minutes of sixth meeting 30th April. PRO HLG 37/62.

—— (1944). *Design of dwellings*. Report of design of dwellings sub-committee of the Central Housing Advisory Committee. (The Dudley Report.) London: HMSO.

—— (1945). *Management of Municipal Housing Estates*. Second report of the Housing Management Sub-Committee of the Central Housing Advisory Committee. (The Balfour Report.) London: HMSO.

Chandran, K. (1993). Presentation Housing Association – the next five years. *Housing Review*, 42 (3), May/June.

CHAR (1988). *Housing for lesbians and gay men*. Conference report.

Chard, C. (1990). A despotic do-gooder. *Financial Times*, 20th January.

Chesler, P. (1973). *Women and madness*. New York: Doubleday.

Churton, A. (1923). *The management of working-class house property*. Substance of paper read for the Rural Housing Association at the Congress of the Royal Sanitary Institute, Hull, 30th July to 4th August. Revised 1925. London: AWHPM.

Clapton, J. (1958). Letter to Society of Housing Managers on behalf of North-East Group. SHM minute book, unpublished.

Clark, H. (1991). *Women, work and stress: new directions*. London: Polytechnic of East London.

Clark, S., Garner, H. et al (eds) (1987). *Our homes: ourselves*. London: Shelter.

Clemens, R. (1992). Letter to M. Brion, 23rd March on behalf of Royal Town Planning Institute.

Clutton, I. (1949). In search of fresh blood. *Junior Organisation Broadsheet*. September.

Coates, K. and Silburn, R. (1980). *Beyond the bulldozer*. Nottingham: Department of Adult Education, University of Nottingham.

Cockburn, C. (1977). *The local state*. London: Pluto Press.

Cockburn, C. (1985). *The machinery of dominance: women, men and technical know-how*. London: Pluto Press.

Cohen, L. and Manion, L. (1985). *Research methods in education*. Revised edition. London: Croom Helm.

College of Estate Management (1936). Advertisement. *SWHEM. Quarterly Bulletin*, 13 (April).

Collison, P. (1963). *The Cutteslowe walls*. London: Faber and Faber.

Commission for Racial Equality (1984). *Hackney housing investigated*. London: CRE.

—— (1990a). *Out of order*. Report of a formal investigation into the London Borough of Southwark. London: CRE.

—— (1990b). *Racial discrimination in an Oldham estate agency*. Report of a formal investigation into Normal Lester and Co. London: CRE.

Committee on Wartime Housing Problems (1943a). Minutes of second meeting, 24th June. PRO HLG 7/1002.

—————— (1943b). Minutes of fifth meeting, 24th September. PRO HLG 7/ 1002 04018.

Connell, R.A. (1987). *Gender and power*. Cambridge: Polity Press.

Cooper, C.L. (ed.) (1982). *Practical approaches to women's career development*. Sheffield: MSC.

Cooper, G.C. and Davidson M. (1982). *High pressure. Working lives of women managers*. London: Fontana.

Cooper, M., Green, M., Payne, J., Riley, D., Scribbins, J., Townsend, J. and Ward, H. (1985). Equal opportunities? Employment in adult education. *NATFHE Journal*, 10 (2), March.

Cooper, S. (1985). *Public housing and private property*. Aldershot: Gower.

Coote, A. and Campbell, B. (1982). *Sweet freedom. The struggle for women's liberation*. London: Pan.

Counter Information Services (1976). *The new technology*. London: CIS.

Cullingworth, J.B. (1960). *Housing needs and planning policy*. London: Routledge & Kegan Paul.

—————— (1963). *Housing in transition*. London: Heinemann.

—————— (1965). *English housing trends* (Occasional Papers in Social Administration No. 13). London: Bell & Sons.

Dallas, L. (1984). Women in housing. *Housing*, 20 (2), February.

Damer, S. and Madigan, R. (1974). The housing investigator. *New Society*, 29 (616) 25th July: 226, 227.

Darke, J. (1984). Women, architects and feminism. In Matrix (1984).

Darley, G. (1990). *Octavia Hill: a life*. London: Constable.

Daunton, M.J. (1983). *House and home in the Victorian city*. London: Edward Arnold.

—————— (ed.) (1984). *Councillors and tenants*. Leicester: Leicester University Press.

Davies, J.G. (1972). *The evangelistic bureaucrat*. London: Tavistock.

Dearlove, J. (1973). *The politics of policy in local government*. Cambridge: Cambridge University Press.

Deaux, K. and Wrightsman, L. (1988). *Social psychology*. Monterey: Brooks/Cole.

De Lyon, H. and Widdowson Nignuolo, F. (1989). *Women teachers: issues and experiences*. Milton Keynes: Open University Press.

Dickson, A. (1982). *A woman in your own right*. London: Quartet.

DOE (1971). *Fair deal for housing*. Cmnd 4728. London: HMSO.

—————— (1977). *Housing policy: A consultative document*. London: HMSO.

Donnison, D.V. (1960). Housing policy since the war (Occasional Papers in Social Administration, No. 1). Welwyn: Codicote Press.

—————— (1967). *The government of housing*. Harmondsworth: Penguin.

Dossett-Davies, J. (1990). Review. *Community Care*, September.

Dowall, J.M.S. (1940). Letter from J.M.S. Dowall, Air Ministry, to H.R. Chapman, Director of Labour, Ministry of Aircraft Production, 25th October. PRO AVIA 15 2508.

Drake, M. (1973). *Applied historical studies*. London: Methuen.

Dresser, M. (1984). Housing policy in Bristol, 1919–30. In Daunton (1984), pp. 156–209.

Dubois, E. (1980). Politics and culture in women's history: contribution to a symposium. *Feminist Studies*, 6 (1) Spring: 28–35.

Dubsky, B. (1982). Letter and committee paper extract 9 November: held by M. Brion.

Dunleavy, P. (1981). *The politics of mass housing in Britain 1945–1975*. Oxford: Clarendon Press.

D.W. (1944). Annual general meeting 20 and 21 November, 1943, *SWHM, Bulletin*, 40 (January): 2–7.

Education and Training for Housing Work Project (1977). *Housing staff*. London: The City University.

Eisenstein, H. (1984). *Contemporary feminist thought*. London: Unwin.

——— (1990). Femocrats, official feminism and the uses of power. In Watson (1990).

Ekstrom, R. Intervention strategies to reduce sex-role stereotyping in education. In Hartnett et al (1979), pp. 217–233.

Ellis, J. (1973). Plumbing the depths. *Chartered Surveyor*, 106 (December): 142.

'E.M.' (1934). Housing estate management in North Kensington. *SWHEM Quarterly Bulletin*, 4 (January).

——— (1939). The Octavia Hill Scholarship Fund. *SWHM. Quarterly Bulletin*, 26 (July): 4–5.

English, J. (1982). The choice for council housing. In *The future of council housing*, ed. J. English. London: Croom Helm, pp. 181–195.

EOC (1982). Sixth Annual Report. Manchester: EOC.

——— (1985a). Wise campaign success. *Equality Now*, 4 (Winter).

——— (1985b). *Women and men in Britain: A statistical profile*. London: HMSO.

EOC and Somerset Council (1982). Access to craft subjects in secondary schools. *EOC Research Bulletin*, 6 (Spring): 73–76.

Epstein, C.F. and Coser, R.L. (1981). *Access to power*. London: Allen and Unwin.

Equal Opportunities Review (1990). Women in the construction industry. *Equal Opportunities Review*, 30 (March/April).

Erikson, K. (1973). Sociology and the historical perspective. In Drake (1973), pp. 13–30.

Ernst, S. and Maguire, M. (1987). *Living with the sphinx*. London: The Women's Press.

Esher, L. (1981). *A broken wave: the rebuilding of England*. London: Allen Lane.

Etzioni, A. (1969). *The semi professions and their organisations*. New York: Free Press.

Fairweather, H. (1976). Sex differences in cognition. *Cognition*, 4: 231–280. Referred to in Griffiths & Saraga (1979).

Felis Domesticus (1936). What women want. Letter to editor. *London Town*, 37 (441), September: 313.

Fenter, F.M. (1960). *Copec adventure*. Birmingham: Birmingham Copec House Improvement Society.

Fenter, M. (1961). Early days in Birmingham. *Society of Housing Managers Quarterly Bulletin*, 5 (4), October.

Figes, E. (1970). *Patriarchal attitudes*. London: Faber & Faber.

Firestone, S. (1980). *The dialectic of sex*. New York: Morrow.

Fogarty, M., Allan, I. and Walters, P. (1981). *Women in top jobs*. London: Heinemann.

Foot, M. (1975). *Aneurin Bevan*. St Albans: Paladin.

Fowler, R.J. (1936). The management of housing estates. *London Town*, 37 February, no 434: 50, 51.

Fox, D. (1973). Housing aid and advice. Unpublished paper: Held by M. Brion.

——— (1974). Adviser on housing management. Comments on the post. Unpublished paper.

France, E. (1935). Will new municipal housing estates become slums? *Municipal Journal*, 44 (18th October): 1881.

Franklin, P. (1976). Presidential address. *Chartered Surveyor*, 109 (December): 135–140.

Franks, C. and Rothblum, E. (eds) (1983). *The stereotyping of women. Its effects on mental health*. New York: Springer.

Fransella, F. and Frost, K. (1977). *Women on being a women*. London: Tavistock.

Freedman, E. (1979). Female institution building and American feminism 1870–1930. *Feminist Studies*, 5 (3) Fall: 512–526.

Freeman, J. (ed.) (1979a). How to discriminate against women without really trying. In Freeman (1979b), pp. 217–312.

——— (ed.) (1979b). *Women – A feminist perspective*. Mayfield: California; Mayfield.

Friedan, B. (1975). *The feminine mystique*. Harmondsworth: Penguin.

——— (1983). *The second stage*. London: Sphere. First published in Britain by Michael Joseph 1982.

Gallagher, P. (1981). Letter to M. Brion, 6th November.

Galton, M. (1926). Housing of the very poor. Paper read in Section D 'Personal and domestic hygiene' of the Congress of the Royal Sanitary Institute, London, 5th to 10th July. London: AWHPM.

Galton, M. (1959). Miss A.E. Dicken: An appreciation. *SHM Quarterly Bulletin*, 4 (14), April: 12.

Garlick, J. (1958). Letter to Society of Housing Managers on behalf of North-West Group, 7th August. SHM minute book unpublished.

Garrett, S. (1987). *Gender*. London: Tavistock.

Gaskell, G. and Sealy, P. (1976). Block 13 Groups: Social Sciences: a third level course: social psychology. Milton Keynes: Open University Press.

Gauldie, E. (1974). *Cruel habitations*. London: Allen and Unwin.

Geldard. (1923). The management of working-class property in a rural district. Paper read at the annual general meeting of the Rural Housing Association. London: Rural Housing Association. Reprinted from *The Clerk of Works Journal*, September.

George, M. (1992). Relatives who are forced to count the cost of caring. *The Guardian*, 16th June.

Gilroy, R. and Woods, R. (1994). *Housing women*. London: Routledge.

Glastonbury, B. (1971). *Homeless near a thousand homes*. London: Allen & Unwin.

GLC's Women's Committee. (1986). Black and ethnic minority women. *Bulletin*, 27 (March).

Goddard, A. (1920). Letter to the editor. *Housing*, 28 (2), 2nd August: 26.

Gold, M. (1934). The second annual provincial conference. *SWHEM. Quarterly Bulletin*, 5 (April).

Goldberg, P. (1968). Are some women prejudiced against women? *Transaction*, 5 (April): 28–30.

Goodman, T. (1992). CSW/Chartwell fund to aid female surveyors. *Chartered Surveyor Weekly*, 23rd July.

Greed, C. (1985). Letter to P. McGurk, Institute of Housing, 3rd May: held by M. Brion.

—— (1991). *Surveying sisters. Women in a traditional male profession*. London: Routledge.

Greve, J., Page, D. and Greve, S. (1971). *Homelessness in London*. Edinburgh: Scottish Academic Press.

Griffith, J.A.G. (1966). *Central departments and local authorities*. London: Allen & Unwin.

Griffiths, P. (1975). *Homes fit for heroes*. London: Shelter.

Griffiths, D. and Saraga, E. (1979). Sex differences in cognitive abilities: a sterile field of enquiry? In: Hartnett et al (1979), pp. 17–49.

Hakim, C. (1979). *Occupational segregation* (Department of Employment Research Paper No. 9). London: Department of Employment.

Halmos, P. (1973). *Professionalism and social change*. Keele: University of Keele.

Hamilton, E. (1941). Planning: housing in the future. *SWHM Quarterly Bulletin*, 34 (October): 4–7.

—— (1948). Letter from E. Hamilton to Miss J. Upcott, 18 April, unpublished, Miss Upcott's Papers: held by M. Brion.

Hancock, W.K. and Gowing, M.M. (1949). *British war economy*. London: HMSO.

Hankinson, A. (1913). The housing problem. Reprinted from the *Manchester Courier*, 12th & 19th July. Printed by Balshaws, Printers, 18 Kingsway, Altrincham (Miss Upcott's papers).

—— (1918). *Cottage property management by trained women*. Reprinted from the *Woman Citizen* (the monthly news sheet of the Manchester and Salford Women Citizens' Association). London: AWHPM.

—— (1919). Women managers. *Housing*, 11 (1) 8th December.

Hansard Society (1990). *Women at the Top*. London: The Hansard Society.

Hargreaves, D. (1979). Sex roles and creativity. In Hartnett *et al.* (1979).

Hargreaves, K. (1983). Letter to R. Best, 24th March. Held by M. Brion.

Harloe, M., Issacharoff, R. and Minns, R. (1974). *The organisation of housing*. London: Heinemann.

Harrison, T. (1978). *Living through the blitz*. Harmondsworth: Penguin.

Harriss, K. (1989). New alliances: socialist-feminism in the eighties. *Feminist Review*, 31 (Spring): 34–54.

Hartnett, O., Boden, G. and Fuller, M. (eds) (1979). *Sex-role stereotyping*. London: Tavistock.

Harvey, O.J. (1954). An experimental investigation of negative and positive relationships between small informal groups through judgemental indices. Doctoral dissertation, University of Oklahoma.

HERA (1988). *HERA membership. Why you should join*. London: HERA.

'H.G.L.A.' (1939). The Great Wen. *SWHM Quarterly Bulletin*, 26 (July): 13.

Hill, M. (née M. Hurst also known as Peggy Hurst) (1981). Interview with M. Brion, 12th March.

Hill, O. (1866). Cottage property in London. *Fortnightly Review*, November. In Hill (1875).

——— (1873). Letter accompanying the account of donations received for work amongst the poor. In *Letters to fellow workers 1872–1906*. London: James Martin.

——— (1875). *Homes of the London poor*. London: Macmillan.

——— (1884). *Colour, space, and music for the people*. Reprinted from *The Nineteenth Century*, May 1884. London: Kegan Paul.

——— (1889). *Management of houses for the poor* (Charity Organisation Society Occasional Paper No. 7) January.

——— (1900). Letter to my fellow workers. London: Waterlow & Sons (private circulation). Reproduced in Hill (1873).

——— (1903). Letter to fellow workers (private circulation). In Hill (1873).

Hill, W.T. (1956). *Octavia Hill*. London: Hutchinson.

Hillier, J., Davoudi, S. and Healey, P. (1988). *Gender issues in planning education* (Education for Planning Association Discussion Paper No. 1), April.

Hirsch, D. (1984). The long road to royal recognition. *Housing*, December: 3.

Hirst, P.H. (1965). Liberal education and the nature of knowledge. In P.H. Hirst (1975). *Knowledge and the curriculum*. London: Routledge & Kegan Paul.

History Workshop Journal editor (1979). Oral history. *History Workshop Journal*, 8 (Autumn) 1979: i–iii.

History Workshop Group (1985). Review Discussion. In search of the past. *History Workshop Journal*, 20 (Autumn): 175–185.

'HMC' (1942). Temporary housing in post-war reconstruction. *SWHM Quarterly Bulletin*, January: 11–13.

Hoinville, G., Jowell, R. and Associates (1978). *Survey research practice*. London: Heinemann.

Holmans, A.R. (1987). *Housing policy in Britain*. London: Croom Helm.

Hort, I. (1934). The Institute of Housing Administration. *SWHEM. Quarterly Bulletin, 4* (January).

——— (1935). Letter to the editor. *Municipal Journal*, 44 (1st November): 1935.

——— (1941). Wokingham rural district. *SWHM Quarterly Bulletin, 34* (October): 2.

——— (1942). Women's conference on planning and housing, May 28th 1942. *SWHM Bulletin*, September: 5, 6.

Housing and Planning Review (1981). National Housing and Town Planning Council. *Housing and Planning Review*, Summer.

Housing Corporation (1993). *Black and minority ethnic housing association strategy annual review*. (Spring; update July). London: Housing Corporation.

Housing Editor (1993). A future in the balance sheet. *Housing*, May.

Housing Research Group (1981). *Could local authorities be better landlords?* London: The City University.

Housing (War Requirements) Committee (1940). Minutes 11th January & 19th January. PRO HLG 68/6.

Houstoun, P. (1985). Letter to the editor. *Housing*, 21 (2) (February).

Howatt, H. (1987). Women in planning – programme of positive action. *The Planner*, August.

Howes, W. (1937). Letter from W.H. Howes as secretary to the Balfour Sub-committee, to Southampton City Council 31st May. PRO HLG 37/5.

'H.R.T.' (1937). Impressions of Dutch housing. *SWHEM. Quarterly Bulletin*, 19 (October): 3–5.

Hulme Tenants Advisory Committee (1937). Report on the work of the Hulme Tenants' Advisory Committee. PRO HLG 37/6.

Humphries J. (1983). The emancipation of women in the 1970s and 80s. From the latent to the floating. *Capital and Class*, 2 (Summer): 6–28.

——— (1988). Women's employment in restructuring America: the changing experience of women in three recessions. In Rubery (1988).

Hurst, M. (1937). The Octavia Hill system in Cape Town. Copy of a memorandum to the town clerk, City of Cape Town, 1st July, unpublished: held by M. Brion.

ICA (1984). Letter to M. Brion, 23rd October.

——— (1989). *Recruitment in the 1990s*. London: Institute of Chartered Accountants.

Illich, I. (1977). *Disabling professions*. London: Marion Boyars.

ILO (1946). *The war and women's employment*. Montreal: International Labour Office.

Inside Housing Editor (1992). Housing assistant wins equal pay. *Inside Housing*, 9 (44) 13th November.

Institute of Housing (IOH) (1936a). Estates management of houses erected for the working-classes. Evidence given before a special Sub-Committee at the Ministry of Health. PRO HLG 37/5.

——— (1936b). Letter to W. Howes, secretary to the sub-committee CHAC, Ministry of Health. 4th November. PRO HLG 37/5.

——— (1956). Annual Report, 1955–56. London: Institute of Housing.

——— (1976). Annual Report, 1975–76. London: Institute of Housing.

——— (1977). Annual Report, 1976–77. London: Institute of Housing.

——— (1978). Annual Report, 1977–78. London: Institute of Housing.

——— (1991). Review of IOH education programme. Paper presented to Equal opportunities sub-committee, 16th April.

——— (1993). Housing welfare services and the housing revenue account. Institute of Housing response to the DOE consultation paper. Coventry: papers held by IOH.

Institute of Housing and Society of Housing Managers (1963). *Report on unification*. London: Institute of Housing and Society of Housing Managers.

Institute of Housing Equal Opportunities Working Group (1992). Report to November council meeting: IOH.

Institute of Housing Managers (1968). *Year book and list of members*. London: Institute of Housing.

——— (1970). Annual Report, 1969–70. London: Institute of Housing Managers.

——— (1971). Annual Report, 1970–71. London: Institute of Housing Managers.

—————— (1972). Annual Report, 1971–72. London: Institute of Housing Managers.

—————— (1973). Annual Report, 1972–73. London: Institute of Housing Managers.

—————— (1975). Annual Report, 1974–75. London: Institute of Housing Managers.

Institute of Housing Professional Practice and Publications Committee (1984a). Supplemental agenda item 1b. Terms of reference for women in housing group, 1984/85.

—————— (1984b). Supplemental agenda item 12a. Report on women in housing working group.

Institute of Housing Women in Housing Working Group (1983a). Agenda papers for 17th February.

—————— (1983b). Minutes of meeting, 20th September.

Institute of Housing Women in Housing Working Party (1984a). Women and the Institute. *Housing*, 20 (2), February.

—————— (1984b). *Advice to women qualifying*. London: Institute of Housing.

—————— (1984c). *Women in the branches*. London: Institute of Housing.

—————— (1984d). *Retraining for women in housing*. London: Institute of Housing.

—————— (1988). Minutes of meeting, 15th July.

Inter-Departmental Committee on Provision of Information and Assistance after Air Raid Damage (1940). Report. PRO HLG 7/518.

James, I. (1959). Secretary's report to Council, 24th September. SHM minute book unpublished.

Jamous, H. and Lemaine, G. (1962). Compétition entre groups d'inégales ressources: experience dans un cadre naturel, premiers travaux. *Psychologie Française*, 7: 216–222.

Jayaweera, H. (1993). Racial disadvantage and ethnic identity: the experiences of Afro-Caribbean women in a British city. *New Community*, 19(3): 383–406.

Jeffery, M.M. (1929). *House property and estate management on Octavia Hill lines* (Occasional Paper, No. 12, Fifth series). London: Charity Organisation Society.

—————— (1930). *Notes on the Octavia Hill system of house property management*. London: National Housing and Town Planning Council.

—————— (1935). Letter to the editor. *Municipal Journal*, 44 (27th December): 2411.

Jeffery, M.M. and Neville, E. (eds.) (1921). *House property and its managements*. London: Allen & Unwin.

Jenkins, P.M. (1958). Letter to the Society of Housing Managers on behalf of South Wales Group, 14th July. SHM minute book unpublished.

Jevons, R. and Madge, J. (1946). *Housing estates*. Bristol: University of Bristol.

John Bull editor (1919a). Voluntary ladies. *John Bull*, 3rd May.

—————— (1919b). A woman intervenes. *John Bull*, 10th May: 2.

—————— (1919c). Tomfoolery at its highest. Waste in training unwanted women. *John Bull*, 5th July: 4.

Johnson, T.J. (1972). *Professions and power*. London: Macmillan.

Joseph, G. (1983). *Women at work. The British experience.* Oxford: Philip Allen.

Joshi, H. (1984). *Women's participation in paid work: further analysis of the women and employment survey* (Department of Employment Research Paper No. 5). London: Department of Employment.

Joshi, H., Layard, R. and Owens, S. (1981). *Female labour supply in post-war Britain: a cohort approach* (Centre for Labour Economics discussion paper No. 79). London: London School of Economics.

'J.S.' (1933). Sir Stanford Downing. *SWHEM. Quarterly Bulletin*, 11. October.

Julienne, L. (1991). Letter to Mark Lupton, Institute of Housing. 22nd April, unpublished: held by M. Brion.

Kanter, R.M. (1977). *Men and women of the corporation.* New York: Basic Books.

Karn, V., Kemeny, J. and Williams, P. (1985). *Home ownership in the inner city: salvation or despair?* Aldershot: Gower.

Kelly, L. (1992). The contradictions of power for women. Paper presented at National Federation of Housing Associations women's conference, April.

Kemp, P. and Williams, P. (1991). Housing management: a historical perspective. In S. Lowe and D. Hughes (eds) *A new century of social housing.* Leicester: Leicester University Press.

Kilbourn, C. and Lantis, M. (1946). Elements of tenant instability in a war housing project. *American Sociological Review*, 11: 65.

Laborde, G. (1987). *Influencing with integrity.* Palo Alto, Calif.: Syntony.

Laffin, M. (1986). *Professionalism and policy: the role of the professions in the central–local government relationship.* Aldershot: Gower.

Lakoff, R. (1975). *Language and woman's place.* New York: Harper and Row.

Lambert, J., Paris, C. and Blackaby, B. (1978). *Housing policy and the state.* London: Macmillan.

Lamplough (1981). Interview with M. Brion, 23rd March.

Larke, H. (1981). Interview with M. Brion, 12th March.

Lawton, D. (1978). *Theory and practice of curriculum studies.* London: Routledge and Kegan Paul.

LCC (1908). Report of the Housing Manager for the year ended 31st March 1908. (Held in Greater London Record Office.)

Lee, V. (ed.). (1976). *Social relationships* (Open University Personality and Learning Block 7 Part 1). Milton Keynes: Open University Press.

Leevers, K. (1986). *Women at work in housing.* London: HERA.

Lerner, G. (1975). Placing women in history: definitions and challenges. *Feminist Studies*, 3 (1/2): 5–15.

Levison, D. and Atkins, J. (1987). *The key to equality.* London: Institute of Housing.

Liverpool Improved Houses Ltd (1928). *Report of a ten months' experiment on the possibility of improving housing conditions in Liverpool.* Liverpool: Liverpool Improved Houses Ltd.

Local Authority Associations and the London County Council (1943). Report to the Minister of Health on Housing. PRO HLG 7/1002.

Local Government Training Board (1984). Letter to L. Poole, 30th March.

London Housing Association Committee (1973). *Housing Associations in London: 1972/73*. London: National Federation of Housing Associations.

London Town (1936). Employment of married women. *London Town, 37* (440), August: 260,261.

Lummis, T. (1987). *Listening to history*. London: Hutchinson.

McCall, R.B. (1990). Promoting interdisciplinary and faculty-service-provider relations. *American Psychologist*, 45 (December): 1319–1324.

Maccoby, E.E. and Jacklin, C.N. (1975). *The psychology of sex differences*. London: Oxford University Press.

McCulloch, D. (1982). Letter to M. Brion, 23 July.

Macey, J. (1980). Interview with M. Brion, 14th January.

Macey, J.P. and Baker, C.V. (1973). *Housing management*. London: Estates Gazette.

McFarlane, B. (1984). Homes fit for heroines: housing in the twenties. In Matrix (1984), pp. 26–36.

McGwire, S. (1993). Has the clock turned back? *Everywoman*, July.

McKay, F. (1976). Self-help therapy. *Spare Rib*, 48 (July). In O'Sullivan (1987).

McKechnie, S. (1990). Foreword. In Sexty (1990), p. 8.

Mackenzie, M. (1936). Women and housing. Letter to the editor. *London Town*, 37 (435) March: 95.

McKerrow, J. (1990). Review of *Octavia Hill* by G. Darley. *Voluntary Housing*, April.

McKinney, D.W. (1973). *The authoritarian personality studies*. The Hague: Mouton.

McLoughlin, J. (1981). The nuts and bolts of equality. *The Guardian*, 31st March.

Malpass, P. (1982). Octavia Hill. *New Society*, 62 (1042) 4th November: 206–208.

——— (1983). Residualisation and the restructuring of housing tenure. *Housing Review*, March/April.

Malpass, D. and Murie, A. (1982). *Housing policy and practice*. London: Macmillan.

Martin, J., and Roberts, C. (1984). *Women and employment. A lifelong perspective*. London: HMSO.

Marwick, A. (1981). *The nature of history*. London: Macmillan.

Mascall, L.W. (1935a). Letter to the editor. *Municipal Journal*, 44 (22nd November): 2161.

——— (1935b). Letter to the editor. *Municipal Journal*, 44 (13th December): 2317.

Matrix (1984). *Making space. Women and the man-made environment*. London: Pluto Press.

Meade-King, M. (1986). How the BBC needs to move away from the workaholic syndrome and encourage women for top jobs. *The Guardian*, 19th March.

Members at Newcastle Under Lyme (1942). Letter to the editor. *SWHM Bulletin, 36* (May).

Members Employed at Stevenage Development Corporation (1958). The future of the Society, 5 November (Paper submitted to Council). SHM minute book unpublished.

Merrett, S. (1979). *State housing in Britain*. London: Routledge & Kegan Paul.

Metters, C. (1981). Letter, *Community Action*, 53 (March–April).

Miller, M. (1935). Letter to the editor. *Municipal Journal*, 44 (1st November).

Millerson, G. (1964). Dilemmas of professionalism. *New Society*. 4th June.

Ministry of Aircraft Production (1941). Minute sheet describing housing schemes. PRO AVIA 15/3852.

Ministry of Health (1919a). House property management by women. *Housing*, 3 (1), 16th August: 44.

——— (1919b). *Manual on unfit houses and unhealthy areas*. London: HMSO.

——— (1919c). In parliament. *Housing*, 4 (1), 30th August: 52.

——— (1919d). Women managers. Management of buildings in Southwark *Housing*, 8 (1) 25th October: 111.

——— (1919e). Circular 40. London: HMSO.

——— (1920a). Women's Advisory Committee. *Housing*, 13 (1), 5th January: 171.

——— (1920b). The management of property. *Housing*, 27 (2) 19th July: 1,2.

——— (1920c). Property management as a solution to the slum problem. *Housing*, 27(2) 19th July: 5,6.

——— (1920d). *Interim report of the committee appointed by the Ministry of Health to consider and advise on the principles to be followed in dealing with unhealthy areas*. London: HMSO.

——— (1921). Second and final report of unhealthy areas committee. London: HMSO.

——— (1935a). Central Housing Advisory Committee. Minute of appointment. PRO HLG 36/1.

——— (1935b). Letter from 'AGH' to 'Mr Wrigley'. 13th December. PRO HLG 36/3.

——— (1936). Housing conference of women's societies. PRO HLG 36/3.

——— (1938). *Report on the management of municipal housing estates* (Circular 1740). London: Ministry of Health.

——— (1939). *Repair of war damage – housing accommodation* (Circular 1810). London: HMSO.

——— (1940a). Government evacuation scheme, Memo. EV 8. London: HMSO.

——— (1940b). Hackney information bureau. Minute signed by G.V. Dyer, 1st October. PRO HLG 7/518.

——— (1941a). *Report on conditions in reception areas by a committee under the chairmanship of Mr. G. Shakespeare MP*. London: HMSO.

——— (1941b). Administration centres in group 1. PRO HLG 7/518.

——— (1942a). Evacuation and the relief of homeless services. Summary of work of local authorities. PRO HLG 7/515.

——— (1942b). *Handbook on billeting and welfare for the use of Chief Billeting Officer*. London: Ministry of Health.

——— (1943a). Memorandum by the Minister of Health to the War Cabinet. Lord President's Committee, July. Appendix 1. PRO HLG 7/1015.

——— (1943b). Building labour for housing work. Memorandum by the

Minister of Health to the War Cabinet, 30th October. PRO HLG 7/ 1015.

—— (1943c). *Requisitioning for families inadequately housed* (Circular 2845). London: Ministry of Health.

—— (1943d). *General scheme for repair of houses* (Circular 2871). London: Ministry of Health.

—— (1944). *The care of the homeless*. London: HMSO.

—— (1946). *Report for year ended 31st March 1946*. London: HMSO.

—— (1949). *Housing progress*. London: HMSO.

Ministry of Housing (1969). *Council housing purposes procedures and priorities* (The Cullingworth Report). London: HMSO.

Mitchell, J. (1971). *Woman's estate*. Harmondsworth: Penguin.

Moberley Bell, E. (1942). *Octavia Hill*. London: Constable.

Morgan, D. (1992). GP division gets tough on sex discrimination. *Chartered Surveyor Weekly*, 24th September.

Morris, J. (1988). *Freedom to lose: housing policy and people with disabilities* (Shelter briefing). London: Shelter.

—— (1993). Feminism and disability. *Feminist Review*, 43 (Spring).

Morris, R.N. and Mogey, J. (1965). *The Sociology of housing*. London: Routledge & Kegan Paul.

Moser, C.A. and Kalton, G. (1971). *Survey methods in social investigation*. London: Heinemann.

Muchnick, D. (1970). *Urban renewal in Liverpool*. London: Bell.

Municipal Journal. (17.2.33). National housing board. *Municipal Journal*, 42: 209,210.

—— (10.3.33a). Future of housing. *Municipal Journal*, 42: 305.

—— (10.3.33b). Slum clearance. Progress of municipal schemes. *Municipal Journal*, 42: 316.

—— (11.8.33). The Moyne Report. *Municipal Journal*, 42: 99–100.

—— (24.11.33). Women property managers. *Municipal Journal*, 42: 1507.

—— (16.3.34). Slum clearance programmes. *Municipal Journal*, 43: 371.

—— (20.4.34). Successful housing management. *Municipal Journal*, 43: 548.

—— (20.7.34). Housing policy. *Municipal Journal*, 43: 1008.

—— (17.5.35). The week in parliament. *Municipal Journal*, 44: 825 & 828.

—— (6.9.35). Housing management is 'ideally women's work'. *Municipal Journal*, 44: 1611.

—— (18.11.38). Notes and comments. *Municipal Journal*, 47: 2549.

Munro, M. and Smith, S.J. (1989). Gender and housing: broadening the debate. *Housing studies*, 4(1) January: 3–17.

Murie, A. (1985). Speech to Institute of Housing South Eastern branch. *Inside Housing*, 24th May: 2.

Murphy, R.J. (1979). Sex differences in examination performance: do these reflect differences in ability or sex-role stereotypes? In Hartnett *et al.* (1979), pp. 159–167.

NCC (1979). *Soonest mended*. London: NCC.

NEDO and RIPA (1990). *Women managers: the untapped resource*. London: Kogan Page.

Nevitt, A.A. (1966). *Housing taxation and subsidies*. London: Nelson.

Newcomb, T.M. (1947). Autistic hostility and social reality. *Human Relations*, 1: 69–86.

NFHA (1958). Annual Report for 1957. London: NFHA.

—— (1985). *Women in housing employment*. London: NFHA.

—— (1987). *Race . . . still a cause for concern*. London: NFHA.

—— (1989). *Race and housing: employment and training guide*. London: NFHA.

—— (1992). *Equal opportunities in housing associations – are you doing enough?* London: NFHA.

NFHA Women in Housing Standing Group (1988). Minutes of meeting, 10th June: Held by M. Brion.

NFHA Women in Housing Working Party (1985). *Women in housing employment*. London: NFHA.

NFHA Women's Conference (1989). Submitted resolutions. Papers held by M. Brion.

Niner, P. (1975). *Local housing policy and practice*. Birmingham: Centre for Urban and Regional Studies, Birmingham University.

Niven, D. (1979). *The development of housing in Scotland*. London: Croom Helm.

Octavia Hill Association of Philadelphia (1906). *Distinctive features of the Octavia Hill Association of Philadelphia*. Philadelphia: Octavia Hill Association.

Octavia Hill Club Editor (1928). Editorial, *Octavia Hill Club Quarterly* (London), December.

Office for Public Management (1992). *Getting more women to the top*. London: OPM.

O'Sullivan, S. (Ed.) (1987). *Women's health. A ' Spare Rib' reader*. London: Pandora.

Over Forty Association (1983). *A place of her own*. London: Over Forty Association.

Over Forty Association for Women Workers (1981). *Annual Report for the year ended 30 September 1981*. London: Over Forty Association.

Painter, M.J. (1980). Policy co-ordination in the Department of the Environment 1970–76. *Public Administration*, 58 (2), Summer.

Palmer, H. (1936). Letter to the editor. *Municipal Journal*, 45 (10th January): 89.

Park, J. (1984). Management standards. Letter to editor. *Housing*, 21(7), July: 10.

Parker, J. and Dugmore, K. (1977). Race and the allocation of GLC housing – a GLC survey. *New Community*, 6(1): 27 & 41.

Parker Morris (1931). *Memorandum upon property management and slum clearance*. London: National Housing and Town Planning Council.

—— (1941). Letter to Minister of Health re Metropolitan Chief Rehousing Officers' conferences. PRO HLG 7/531.

—— (1943). Letter to the Ministry of Health, 8th September. PRO HLG 1002.

Peach, C. and Byron, M. (1993). Caribbean tenants in council housing: 'race', class and gender. *New Community*, 19(3): 407–423.

Pearce, F. (1936). Congratulations. Letter to the editor. *London Town*, 37(441), September: 314.

Pearson, M. (1992). Sophisticated discrimination. *New Law Journal*, 142 (3rd April): 466–467.

Perry, O. (1990). Policy for women architects. Paper presented to RIBA Council, November.

Phenix, P.H. (1964). *The realms of meaning*. Maidenhead: McGraw-Hill.

Philipp, G.E. (1958). Letter to Society of Housing Managers on behalf of London group, 10th September. SHM minute book unpublished.

Pixner, S. (1978). For therapy. *Spare Rib*, 69 (April). In O'Sullivan, S. (1987).

Pomata, G. (1993). History, particular and universal: on reading some recent women's history textbooks. *Feminist Studies*, 19(1), Spring.

Poole, L. and Porter, J. (1988). *Tackling violence against housing staff*. London: Institute of Housing and NFHA.

Popper, K. (1957, 1986). *The poverty of historicism*. London: Routledge & Kegan Paul.

Portelli, A. (1981). The peculiarities of oral history. *History Workshop Journal*, 12 (Autumn): 97–107.

Porter, R. (1990). Gender menders. *Evening Standard*, 25.1.90.

Povall, M. and Hastings, J. (Eds.) (1982). *Managing or removing the career break*. Sheffield: MSC.

Power, A. (1985). The development of unpopular council housing estates and attempted remedies 1895–1984. Submitted for degree of PhD, July, London School of Economics.

—————— (1987). *Property before people*. London: Allen & Unwin.

Pring, R. (1978). Philosophical issues. In Lawton (1978).

Rackham, N. and Morgan, T. (1977). *Behaviour analysis in training*. Maidenhead: McGraw-Hill.

Rao, N. (1990). *Black women in public housing*. London: Black women and housing group, London race and housing research unit.

Ratcliffe, M. (1939). Housing in Australia. November 1938. *SWHM. Quarterly Bulletin*, Members' supplement, April: 3.

Rathbone, E. (1935). A woman's view of the Local Government Service. *Municipal Journal*, 44 (3rd May): 734.

Ravetz, A. (1974). *Model estate*. London: Croom Helm.

Reid, I. and Wormald, E. (1972). *Sex differences in Britain*. London: Grant McIntyre.

Rendel, M. (Ed.) (1981). *Women: Power and political systems*. London: Croom Helm.

Riggs, F.W. (1990). Interdisciplinary Tower of Babel. *International Social Science Journal*, 126 (November): 577–592.

Riley, D. (1979). War in the nursery. *Feminist Review*, 2: 82–107.

Robarts, S. with Coote, A. and Ball, E. (1981). *Positive action for women*. London: National Council for Civil Liberties.

Roberts, F.C. (1978). *Obituaries from 'The Times' 1971–75*. Reading: Newspaper Archive Developments.

Robertson, G.A. (1958). Letter to Society of Housing Managers on behalf of Scottish group, 8th September. SHM minute book unpublished.

Rose, C. (1985). *Accelerated learning*. Gt Missenden, Bucks: Topaz Publishing.

Rowbotham, S. (1973). *Hidden from history: three hundred years of oppression and the fight against it.* London: Pluto Press.

Rowbotham, S., Segal, L. and Wainwright, H. (1979). *Beyond the fragments. Feminism and the making of socialism.* London: Merlin Press.

Rowe, R.P.P. (1931). *A work of slum reclamation* (reprinted from *The Times*, 8th April 1931. London: Times Publishing Company.

RTPI (1988). *The employment of women in the planning profession.* London: Royal Town Planning Institute.

RTPI North West Branch Women and Planning Working Party (1982). Working within the Royal Town Planning Institute. Workshop report. In *Women and the planned environment.* London: Central London School of Environment Planning Unit.

Rubery, J. (1988). *Women and the recession.* London: Routledge & Kegan Paul.

Rubin, J.Z. and Brown, B.R. (1975). *The social psychology of bargaining and negotiation.* London: Academic Press.

Runnymede Trust (1975). *Race and council housing in London.* London: Runnymede Trust.

———— (1985). *Housing association allocations: achieving racial equality.* London: Runnymede Trust.

Ryan, M.P. (1979). The power of women's networks. *Feminist Studies*, 5 (1): 67–83.

———— (1983). The power of women's networks. *Feminist Studies*, 15 (1), Spring: 6–83.

Ryder, R. (1984). Council house building in County Durham, 1900–39. The local implementation of national policy. In Daunton (1984).

Safilios-Rothschild, C. (1972). *Toward a sociology of women.* Lexington/ Toronto: Xerox College Publishing.

Samuel, R. (1980). On the methods of history workshop: a reply. *History Workshop Journal*, 9 (Spring): 163–175.

Sanderson, A. (1935). Letter to Minister of Health 10th December 1935. PRO HLG 36/3.

Sandle, S.E. (1980). The housing profession: A case study of processes and consequences. Brunel University, unpublished dissertation.

Sayers, J. (1979). On the description of psychological sex differences. In Hartnett et al (1979), pp. 46–56.

Segal, L. (1987). *Is the future female?* London: Virago.

Selbourne, D. (1980). Critique on the methods of history workshop. *History Workshop Journal*, 9 (Spring): 151–159.

Sexty, C. (1990). *Women losing out. Access to housing in Britain today.* London: Shelter.

Sharp, E. (1969). *The Ministry of Housing and Local Government.* London: Allen & Unwin.

Shaw, K.M. (1958). Letter to the Society of Housing Managers, 7th May. SHM minute book unpublished.

Sherif, M. (1967). *Group conflict and co-operation.* London: Routledge & Kegan Paul.

Shibley Hyde, J. and Rosenberg, B.G. (1976). *Half the human experience.* Lexington/Mass.: D.C. Heath.

SHM (1958). Memorandum to the local government examinations board on the list of examinations recognised for promotion purposes. SHM minute book unpublished.

——— (1960). Twenty-third Annual Report, 1959–1960. London: SHM.

——— (1962). Twenty-fifth Annual Report, 1961–1962. London: SHM.

——— (1963). Twenty-sixth Annual Report, 1962–1963. London: SHM.

——— (1964). Twenty-seventh Annual Report, 1963–1964. London: SHM.

SHM Ad Hoc Committee on Recruitment, Training & Membership (1959). Report to council, July. SHM minute book unpublished.

SHM Members at Stevenage (1958). Comments on the future of the society, 9th November. SHM minute book unpublished.

SHM Minutes (17.7.54). Meeting of council. SHM minute book unpublished. Minute books held by Institute of Housing, Coventry.

——— (9.8.54). Notes of a meeting held at the Society's office. SHM minute book unpublished.

——— (11.9.54–13.5.61). Meeting of council. SHM minute book unpublished.

Short, J.R. (1982). *Housing in Britain: the post-war experience.* London: Methuen.

Silverstone, R. (1980). Accountancy. In Silverstone & Ward (1980).

——— (1990). *Recruitment and retention of chartered accountants in England and Wales.* London: ICA recruitment trends study group, January.

Silverstone, R. and Ward, A. (eds) (1980). *Careers of professional women.* London: Croom Helm.

Slizowski, L. (1984). Politics and housing officers. *Housing,* 20 (12).

Smith, B. (1989). *Changing lives: women in European history since 1700.* Lexington, Mass.: D.C. Heath.

Smith, D. and Whalley, A. (1975). *Racial minorities and public housing.* London: PEP.

Smith, H.W. (1975). *Strategies of social research.* London: Prentice-Hall.

Smith, M. (1967). Letter to the editor . . . and even brighter girls. *Chartered Surveyor,* 100(6) December: 282.

——— (1973). Letter to the editor. *Housing,* November: 54.

——— (1974). Letter to M. Brion, 7th May.

——— (1979). Interview, 14th February.

——— (1982). Letter to J. Underwood, 6th November.

Smith, S. (1986). *Separate tables? An investigation into single-sex setting in mathematics.* London: HMSO.

Snook, M.W.G. (1944). Annual provincial conference, Hereford. *SWHM Bulletin, 41* (June): 2–4.

Solomos, J. (1991). The politics of race and housing. *Policy & Politics,* 91 (July): 147–157.

Somerville, J. (1982). Women, a reserve army of labour? *M/F,* 7 (35): 60.

South Bank Polytechnic (1987). Women and their built environment. Conference in the Faculty of the Built Environment, South Bank Polytechnic, London.

Southwark, M.B. (1942). Memorandum upon the difficulties encountered by the Southwark Metropolitan Borough Council with respect to repair and maintenance of premises requisitioned for rehousing purposes for which *inclusive* rentals are authorised by the Ministry of Health. PRO HLG 7/ 511.

268 Bibliography and references

Special Committee of the Women's Housing Conference (1936). Memorandum submitted to the Minister of Health. PRO HLG 36/3.

Spender, D. (1980). *Man made language*. London: Routledge & Kegan Paul.

Spicker, P. (1985). Legacy of Octavia Hill. *Housing*, June: 39–40.

Stacey, M. and Price, M. (1980a). *Women, power and politics*. London: Tavistock.

——— (1980b). Women and power. *Feminist Review*, 33: 5.

Stanford, T. (1989). Letter from Tony Stanford, Chief officer training and personnel, NFHA to M. Brion, 18th January.

Stott, M. (1984). A woman's place is in housing. *The Guardian*, 26th June.

Strange, K.H. (1941). Letter to the editor. *SWHM Quarterly Bulletin*, October: 9.

Sturge, M.D. (1944). Annual general meeting. *SWHM Bulletin*, 42 (December): 3–5.

'Subject' (1936). The object of interference. Letter to the editor. *London Town*, 37 (441), September: 314.

Sutcliffe, A. (ed.) (1974). *Multi-storey living. The British working class experience*. London: Croom Helm.

SWHEM (1933). First Annual Report. London: SWHEM.

——— (1934a). *Housing estate management by women*. London: SWHEM.

——— (1934b). Annual Report for 1933–34. London: SWHEM.

——— (undated). Training scheme. As the Society's address is given as Suffolk Street, this training scheme was probably issued around 1935–36. London: SWHEM.

——— (1936a). Evidence presented to the CHAC Sub-Committee on housing associations and housing management. Paper headed 'recent totals'. PRO HLG 37/5.

——— (1936b). Annual Report for 1935–36. London: SWHEM.

——— (1937). Annual Report for 1936–37. London: SWHEM.

——— (1938). Annual Report for 1937–38. London: SWHEM.

——— (1939). Annual Report for 1938–39. London: SWHEM.

SWHEM Minutes (22.1.33–13.2.37). Report of executive to council on proceedings since 22nd January 1933. SWHEM minute book unpublished.

SWHEM Quarterly Bulletin. (April 1933). Editorial. *SWHEM. Quarterly Bulletin*, 1.

——— (July 1933a). The departmental committee on housing. *SWHEM Quarterly Bulletin*.

——— (July 1933b). Appointments. *SWHEM Quarterly Bulletin*, 2.

——— (October 1933a). Housing in Western Germany. *SWHEM. Quarterly Bulletin*, 3.

——— (October 1933b). Members' news. *SWHEM. Quarterly Bulletin*, 3.

——— (October 1933c). Report of the departmental committee on housing. *SWHEM Quarterly Bulletin*, 3.

——— (April 1934a). Addresses. *SWHEM. Quarterly Bulletin*, 5.

——— (April 1934b). Appointments. *SWHEM. Quarterly Bulletin*, 5.

——— (July 1934a). Appointments. *SWHEM. Quarterly Bulletin*, 6.

——— (July 1934b). Notes. *SWHEM. Quarterly Bulletin*, 6.

——— (July 1934c). Sir George Duckworth. *SWHEM. Quarterly Bulletin*, 6.

——— (July 1934d). Working-class dwellings in and around Paris. *SWHEM. Quarterly Bulletin*, 6.

―――― (July 1934e). Tours to Copenhagen and Holland. *SWHEM. Quarterly Bulletin*, 6.

―――― (April 1935). Members' news. *SWHEM Quarterly Bulletin*, 9.

―――― (July 1935). Address by Miss Samuel on her USA tour. *SWHEM. Quarterly Bulletin*, 10.

―――― (October 1935a). International congress for scientific management. *SWHEM. Quarterly Bulletin*, 11.

―――― (October 1935b). Institute of Housing Administration. *SWHEM. Quarterly Bulletin*, 11.

―――― (October 1935c). Members' news and notes. *SWHEM. Quarterly Bulletin*, 11.

―――― (January 1936a). Annual general meeting. *SWHEM. Quarterly Bulletin*, 12.

―――― (January 1936b). Report of an address by Mr Parker Morris. *SWHEM. Quarterly Bulletin*, 12.

―――― (April 1936). College of Estate Management advertisement. *SWHEM. Quarterly Bulletin*, 26.

―――― (October 1936). Women's housing conference. *SWHEM. Quarterly Bulletin*, 15.

―――― (January 1937). National Housing and Town Planning Council. *SWHEM. Quarterly Bulletin*, 16: 9.

―――― (April 1937). Addresses by members. *SWHEM. Quarterly Bulletin*, 17.

SWHM (1937). Memorandum and Articles of Association. London: SWHM.

―――― (1938). First Annual Report, 1937–38. London: SWHM. (Reports held by M. Brion.)

―――― (1939). Second Annual Report, 1938–39. London: SWHM.

―――― (1940). Third Annual Report. London: SWHM.

―――― (1941). Fourth Annual Report. London: SWHM.

―――― (1942). Fifth Annual Report. London: SWHM.

―――― (1943a). Sixth Annual Report. London: SWHM.

―――― (1943b). Memorandum drawn up at the request of the sub-committee of the central housing advisory committee of the Ministry of Health on the design of dwellings. London: SWHM.

―――― (1944). Seventh Annual Report. London: SWHM.

―――― (1945). Eighth Annual Report. London: SWHM.

―――― (1946). SWHM Annual Report, 1945–1946. London: SWHM.

―――― (1948). SWHM Annual Report, 1947–1948. London: SWHM.

SWHM Bulletin (May 1942). Notes on council proceedings. *SWHM Bulletin*, 36: 1.

―――― (September 1942). Notes on council proceedings. *SWHM Bulletin*, 37: 1.

―――― (January 1943). Annual general meeting. *SWHM Bulletin*, 38, January.

―――― (June 1943a). Notes on council proceedings. *SWHM Bulletin*, 39, June: 2.

―――― (June 1943b). Notes on council proceedings. *SWHM Bulletin*, 39, June 1943: 1.

―――― (January 1944). Annual general meeting. *SWHM Bulletin*, 40: 2–7.

SWHM Minutes (17.6.39–8.12.56). Minutes of council. SWHM minute book unpublished. (Minute books held by Institute of Housing, Coventry.)
———— (30.9.39). Meeting of emergency executive. SWHM minute book unpublished.
———— (5.11.40). Proceedings of emergency executive, 14.9.40–5.11.50. SWHM minute book unpublished.
———— (20.12.46/11.1.47). Minutes of ad hoc committee on admission of men. SWHM minute book unpublished.
———— (September 1947). Ad hoc committee on admission of men. SWHM minute book unpublished.
———— (14.2.48) Meeting of council. SWHM Minute Book unpublished.
SWHM. *Quarterly Bulletin* (January 1938a). First general meeting. *SWHM. Quarterly Bulletin*, 20: 4–9. (*Quarterly Bulletins*: papers held by M. Brion.)
———— (January 1938b). The Housing Centre. *SWHM. Quarterly Bulletin* 20: 17.
———— (July 1938). The annual provincial conference. *SWHM Quarterly Bulletin*, 22: 4–5.
———— (April 1939a). Proceedings of council, January to March 1939. *SWHM. Quarterly Bulletin*, 25. Members' Supplement: 1.
———— (April 1939b). Problems of property management in time of war. *SWHM. Quarterly Bulletin*, 25: 7.
———— (July 1939). Social assistance. *SWHM. Quarterly Bulletin* 26: 16,17.
———— (January 1940). Annual general meeting. *SWHM. Quarterly Bulletin*, 28: 4.
———— (January 1941). Notes on council proceedings. *SWHM. Quarterly Bulletin*, 32: 1.
———— (January 1942). The reconstruction movement and a national plan. Report of meeting addressed by Professor Holford, 14 October 1941. *SWHM. Quarterly Bulletin*: 7–11.
Tabor, M.E. (1927). *Octavia Hill*. London: Sheldon Press.
Tajfel, H. (1981). *Human groups and social categories*. Cambridge: Cambridge University Press.
Tarn, J.N. (1973). *Five per cent philanthrophy*. Cambridge: Cambridge University Press.
The Times. (1934). To 'blaze a new trail'. City's Manageress of Housing. *The Times*, 8th October. Cutting supplied by M. Hurst.
———— (1939). Careers for girls. Women property managers. *The Times*, 20th January.
Thompson, J. (1931). The administration of muncipial housing estates. London: Paper presented to the Institute of Public Administration, February.
———— (1935a). *A day in my official life: housing estate manager*. London: Institute of Public Administration.
———— (1935b). Letter to the editor. *Municipal Journal*, 44 (29th November): 2211.
———— (1936). Memorandum on the detailed working of the Octavia Hill system in the County Borough of Rotherham. PRO HLG 37/5.
———— (1938). Octavia Hill comes to Southall. *SWHM. Quarterly Bulletin*, 20 (January): 12–14.

Thompson, J.M. (1928). Conference of municipal housing managers. *Octavia Hill Club Quarterly*, December: 8,9.

——— (1931). *The administration of municipal housing estates*. London: AWHPM.

Thrupp, B.A. (1929). *The scope of a house property management department*. Paper presented at sessional meeting of the Royal Sanitary Institute, Shrewsbury, 12th October. London: SWHPM.

——— (1932). *Octavia Hill methods of managing property*. Chartered Surveyors Institution, Lancashire and Cheshire County Branch.

——— (1947). A housing manager on her contacts with Whitehall and council chambers. *Society of Women Housing Managers' Quarterly Bulletin*, January: 1–2.

Titmuss, R.M. (1950). *Problems of social policy*. London: Croom Helm.

Todd, M. and Karn, V. (1993). Matters of strategic importance. *Inside Housing*, 10 (28), 23rd July.

Tosh, J. (1984). *The pursuit of history*. London: Longman.

Townroe, B.S. (1934). Future housing policy. *Municipal Journal*, 43 (27th July): 1042.

Toynbee, P. (1984). There is no hiding place . . . *The Guardian*, 10 December: 10.

Tucker, J. (1966). *Honourable estates*. London: Gollancz.

Una e Minimus (1936). Some fallacies. Letter to the editor. *London Town*, 37 (41) September: 313,314.

Ungerson, C. and Karn, V. (1980). *The consumer experience of housing*. Aldershot: Gower.

Upcott, J.M. (1918). Management and housing. Reprinted by kind permission of the *Economist*. London: AWHPM.

——— (1920). Letter to "Katherine" from No. 10 Government Hostel, Dudley, 4th January. Miss Upcott's papers. Held by M. Brion.

——— (1923). Women house property managers. *Building News*.

——— (1927a). New career for women. Letter to the *Morning Post*, 8th February. Dated by hand by Miss Upcott.

——— (1927b). *Memorandum upon the management of a municipal estate at Chesterfield on the lines initiated by the late Miss Octavia Hill*. London: National Housing and Town Planning Council.

——— (1928). Management of municipal housing estates on Octavia Hill lines. (Outline for talk given at Plymouth). 18th July. Miss Upcott's papers. Held by M. Brion.

——— (1932). Women house property managers, 1932. Reprinted from the *Woman's Leader*, July.

——— (1933). Typewritten note of dissolution of AWHPM and handover of funds to SWHEM. London: AWHPM.

——— (1962). Fellow-workers 1912–1932. *Society of Housing Managers Quarterly Bulletin*, 5 (8), October: 5–7.

——— (1979). Interview with M. Brion, 17th February.

Vallance, E. (1979). *Women in the house*. London: Athlone Press.

Walker, D. (1983). Now we shall all pay the rent for the council house poor. *The Times*, 2nd February.

Walker, J. (1988). Women, the state and the family in Britain: Thatcher economics and the experience of women. In Rubery (1988).

Wallach Scott, J. (1983). Women in history. *Past and Present*, December: 125–157.

Walsh, M. (1934). To blaze a new trail – City's manageress of housing. *The Times*, 8th October.

Walton, R.G. (1975). *Women in social work*. London: Routledge & Kegan Paul.

Waring, M. (1989). *If women counted*. London: Macmillan.

Wartime Housing Problems Committee (1944). Minutes of sixth meeting, 7th January 1944. PRO HLG 7/1015.

Watson, S. (Ed.) (1990). *Playing the state*. London: Verso.

Watzlawick, P., Weakland, J. and Fisch, R. (1974). *Change. Principles of problem formation and problem resolution*. London: W. W. Norton.

Weinreich-Haste, H.J. (1979). What sex is science? In Hartnett et al (1979), pp. 168–181.

Welch, K. (1986). The black women and housing group. *GLC Women's Committee Bulletin*, March.

Who's Who. (1934). London: A. & C. Black.

Wigfall, V. (1980). Architecture. In *Careers of professional women*. London: Croom Helm.

Wilensky, H.L. (1964). The professionalization of everyone. *American Journal of Sociology*, 70 (2), September: 143.

Wilson, E. (1980). *Only halfway to paradise*. London: Tavistock.

Wilson, R. (1963). *Difficult housing estates*. London: Tavistock.

Wohl, A.S. (1977). *The eternal slum*. London: Edward Arnold.

Wolpe, A.M. (1977). *Some processes in sexist education*. London: WRRC Publications.

Women and Housing Group (1980a). Programme for women and housing conference. Papers held by M. Brion.

—— (1980b). Women and housing course programme. Papers held by M. Brion.

—— (1984). Older women. Extract from submission to GLC women's committee. *Older Feminists' Network newsletter*, 13. November.

Women and Manual Trades (1988). Training at Hackney DLO. *WAMT Newsletter*, March.

Women in Architecture Sub-group (1984). Report to the membership committee of RIBA. September.

Women in Housing Group, London (1981a). Minutes of first meeting, 19th May.

—— (1981b). *Newsletter*, November.

—— (1984). *Newsletter*, September. Conference Programme 29 September. Papers held by author.

Women in Housing Group, Sheffield (1984). Conference programme, 29th September. Papers held by M. Brion.

Women's Design Service (1992). *Race and gender. Architectural education*. London: Women's Design Service.

Women's Group on Public Welfare (1943). *Our towns*.

Women's Pioneer Housing Ltd (1935). Publicity leaflet. London: Women's Pioneer Housing Ltd.

Woodall, J., Showstack, A., Towers, B. and McNally, C. (1985). Don't tell me the old boys' story. *The Guardian*, 10th September.

Yorke, H. and Lumsden, (undated, around 1910). *Miss Octavia Hill's method of house management* (printed pamphlet). Clearly written after Octavia Hill's death, but possibly before the formation of an association.

Young, K. and Kramer, J. (1978). *Strategy and conflict in metropolitan housing*. London: Heinemann.

Zemon Davis, N. (1976). Women's history in transition: the European case. *Feminist Studies*, 3(2), Spring/Summer: 83–103.

Index

In this index, authors are indicated by surname and initials (e.g. Brion, M.). People involved in housing management are either given full names (e.g. Hill, Octavia) or a title (e.g. Fenter, Miss E.M.). Page numbers in *italic* indicate a Table and in **bold** a Figure.